IEE TELECOMMUNICATIONS SERIES 34

Series Editors: Professor J. E. Flood
              Professor C. J. Hughes
              Professor J. D. Parsons

Digital broadcasting

Other volumes in this series:

Digital broadcasting

P. Dambacher

The Institution of Electrical Engineers

Originally published in German by R. v. Decker, 1994

English edition published by: The Institution of Electrical Engineers,
London, United Kingdom

English edition © 1996: The Institution of Electrical Engineers

British Library Cataloguing in Publication Data

A CIP catalogue record for this book
is available from the British Library

ISBN 0 85296 873 6

Printed in England by Short Run Press Ltd., Exeter

Contents

Introduction

Sound and television broadcasting are undergoing a revolutionary development from the established analogue to digital technologies, especially digital signal processing. This change has become evident over the past ten years, taking shape in many development areas (including the audio and in particular the video baseband signal) to be largely completed within the next decade.

This means leaving the path of developing the newer and latest technology to be compatible with the existing. Technical development requires a great amount of engineering resources and every technical process, for example PAL to Palplus at present, needs equipment for further use of the existing technique. This dual approach takes away the efficiency of the engineering resources, in contrast to applying all efforts into an independent and revolutionary (digital) process. On the other hand, the new digital technology, being independent and not relying on the existing systems, allows for a greater degree of innovation, which however, must be demonstrated to educate the consumer into buying the new technology.

Adopting digital technologies requires conversion of the present analogue signals into datastreams with two defined states: the binary states 1 and 0 [1]. Prior to digital processing, the analogue signals have to be transformed into discontinuous discrete quantities by the sample and hold process in which there is a transformation of a linear curve into an equivalent 'staircase' The digital quantity is a combination of a time-discrete value and a numerical value. The digital value corresponding to a level of the 'staircase' has an explicit value at a defined timepoint.

To generate a numerical sequence, the analogue signal must first be sampled, i.e. instantaneous values are measured on a discrete time scale with constant time slots (e.g. 1 µs). Using quantisation, numbers between 0 and a defined maximum value are assigned to the measured values. The sampled value so derived is coded by a binary number. Apart from the binary system, any other number system may of course also be used.

The digital technique has the following distinct advantages:

- Any signal quality can be represented, the quality can for instance be doubled by increasing the accuracy by 1 bit.
- The digital signal transmission is less insensitive to interference such as white noise or discrete frequencies.
- With digital representation, the values can easily be manipulated, processed and stored.
- Through the use of channel coding allowing error detection and correction, the transmission reliability of digital signals can be adapted to the particular channel used, e.g. terrestrial transmission, satellite or cable transmission.

An essential condition in the advance to digital technologies is realtime signal processing. It was only in the last few years this condition could be met through the development of faster digital circuits and special computer architectures (microcomputers, programmable signal processors). The era of digital signal processing (DSP) has begun taking a firm foothold in sound and television broadcasting.

The components of DSP techniques are:

- The **analogue/digital convertor** provides discrete-value levels for the DSP system. With noncritical sampling rates, as in audio techniques (around 20 µs), quantisation is made by means of successive approximation. With this method, the sampling time is divided into sample periods, during which the output value is gradually approaching the threshold value at the input. This is a special A/D conversion where the digital value is generated in a stepping process with a D/A convertor and a comparator. At every step the digital output is increased, D/A converted and compared with the input until the output and input values are equal. For sampling rates that are much higher, say in the order of 50 ns, as required by video techniques, fast parallel convertors have to be used. Using the Nyquist sampling theorem, the signals are passed through a lowpass filter prior to their A/D conversion, the so-called antialiasing filter, with a cutoff frequency of half the sampling frequency.
- **Digital filters** play an important part in DSP. They feature such essential benefits as freely selectable transfer characteristics, avoiding problems associated with component tolerances and unlimited stability. Distinction is made between transversal filters without feedback and recursive filters where the output signal is coupled to the input in a specific way. Digital filter elements include adders, multipliers and delay elements in the form of shift registers.
- Processing of the signals takes place in the **signal processors** with so-called hardwired algorithms or with user-programmable firmware. The programmable signal processor behaves like a microcomputer, which means that command sequences with conditional and unconditional jumps can be performed. In addition to the jumps, the main functions are multiplication and accumulation.

- The **digital/analogue convertors** are relatively noncritical and they enable the return to the analogue condition. As with A/D conversion, the signal must pass through a lowpass filter with a cutoff frequency of half the sampling frequency after the D/A convertor.
- DSP is performed within a **defined clock period,** which makes the time domain discrete. All process operations are controlled and synchronised. The system clock is crystal stabilised and used to derive all required clock frequencies.

In addition to the above outlined general advantages of digital techniques, DSP also provides other special advantages:

The sound and TV broadcasting equipment can largely be manufactured without any adjustment presets. As the adjustment of digital circuits is unnecessary, the transfer characteristics of analogue components can be stored and aligned digitally. The use of software makes the flexibility of digital equipment much greater than analogue equipment.

The firmware/software provides another important asset: although the development costs are high, the cost of reproduction of software is negligible (just the duplication of EPROMs or discs) so the unit cost is effectively equal to the development cost divided by the number of pieces. The larger the quantity of equipment produced, the more negligible are the software costs.

The term sound and television broadcasting used in this book refers to the whole world of sound broadcasting from the microphone through to the loudspeaker, and the whole world of television from the TV camera through to the screen of the domestic TV set. This means that topics such as sound broadcasting and TV studios, the infrastructure of the transmission links using terrestrial transmitters, satellites and broadband cables, and the consumer/receiver equipment will all be dealt with.

This book aims give the reader an impression and overview of sound broadcasting and television standards as well as the technical and technological achievements in these fields over the next ten years. Selected chapters deal with measurements and monitoring.

The book can therefore be divided into three main sections:

- Innovations in sound broadcasting and television in the past ten years
- Plans for the coming decade
- Author's contributions in the form of discussions and solutions.

To describe the past decade is easy, to predict the next, one has to adopt a certain approach taking into account previous developments and, with the knowledge of the present standardisation work of international committees as well as of the research and planning work in universities and institutes, the next ten years can be projected by way of extrapolation.

In making any forecasts for the future, 'theme sensing' – keeping track of new trends in development – is an important aspect. The so-called filter syndrome, the subjective view of objective 'facts' (truth), resulting in seeing only those arguments leading to digital transmission, cannot of course be fully eliminated. Arguments supporting one's own vision will automatically be given preference.

Another possible approach to technical forecasts can best be described by the slogan **'Technology Push – Market Pull'**. In other words, everything that is technologically possible and feasible is required and accepted by the market. Considering the progress in LSI (large scale integration) technology, there is a strong argument for complete renewal of sound and TV broadcasting.

The revolution in integration techniques and microelectronics has made the sound and TV broadcasting reforms possible. New dimensions were in particular reached by functional density, reliability and speed of the ICs. Prices dropped, for storage elements for instance by a factor of 1000. DRAM capacity (dynamic random access memory), which in 1980 was as low as 64 kbyte, will reach 1/4 Gbyte (Table I.1) around the turn of the century [1]. The integration density of logic circuits has shown a similar development, and as computing power is a function of the integration density multiplied with the speed, progress will become even faster.

The keywords are:

- ASIC (application specific integrated circuit)
- Transputer (processors for parallel processing)
- RISC computer (reduced instruction set computing) or
- Submicron logic processors (CMOS: complementary metal-oxide semiconductor).

	1980	1990	2000
Dynamic RAM (DRAM)	64 kbit	4 Mbit	256 Mbit
Structure width		1 μm	0.3 μm
Gates per mm^2	10^3	10^4	5×10^4
Switching speed in ns	10	0.5	0.1
Functional elements per chip	10^5	10^7	5×10^8

Table I.1: Development of chip integration [1]

There has also been a change in large-signal technologies. Solid-state amplifiers up to 10 (20) kW in bands I to IV/V have become a true alternative to the vacuum technology using tetrodes and klystrons; comparative assessments being based on the cost of ownership.

Finally, there is one more term used in technical and technological forecasts: the technology *S*-curve. This is a plot of the development, or the performance level of a development or a technology against time, resulting in an *S*-curve (Fig. I.1). At the start, the curve shows a rather slow development and penetration of the market. The central part shows a stabilisation of the basic technology, with continuous improvement and a market-related linear growth. At the end of its life, the technology is fully developed and the curve flattens again: the market is saturated and equipment replacement is the only business available (reinvestment). For example, for TV transmitters with a tetrode, the technology is well proven and the terrestrial transmitter market is saturated; it is time for a new technology, namely solid state. A new technology *S*-curve must begin, and at this phase of radical change is where sound and television broadcasting is today.

What did this change bring about?

In the field of sound broadcasting it certainly was the compact disc in the mid-1980s, which satisfied the 'golden ear'. Later in the 1990s when CD technology could be bought at prices acceptable to the consumer, a new quality consciousness was born of which sound broadcasting technology had to take account and continues to do so. Digital audio broadcasting via satellite and cable and the development of a terrestrial digital audio broadcasting system are the result.

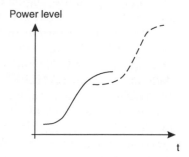

Fig. I.1: Technology S-curve

Regarding television, it is not so easy to see an obvious reason for the change. There is no TV equivalent to CD technology.

Is there dissatisfaction with the quality of the PAL, SECAM and NTSC picture? (PAL: phase alternation line; SECAM: séquentielle à mémoire; NTSC: national television system committee). Has the end of the technology curve been reached or are there new possibilities for LSI technology?

Without going into details, it has to be said that television is undergoing a great change. The keywords for the new television are: 16:9 format and HDTV (high definition television). In sound broadcasting there was and is continuous movement towards entirely digital transmission. In TV there was an evolutionary way via digital sound (and data) and analogue vision, e.g. PAL plus, D2MAC, HDMAC, and also a purely digital method: digital video coding.

According to experts, the development of the digital line in Europe is about five years behind the analogue, whereas in the United States, it is in the testing and field trial phase. In the US, there was no intention to keep analogue video. The trend goes directly from NTSC to full digital transmission, not via MAC but straight to HDTV via the terrestrial transmitter network, with TV-channel compatibility (6 MHz bandwidth per TV channel). In addition to the aim of achieving higher quality, there is the alternative of broadcasting several programmes per channel.

The move to digital techniques in sound and television broadcasting brings, of course, a number of further advantages such as higher immunity to interference, possibility of regeneration, use of digital modulation methods such as PSK (phase shift keying) or QAM (quadrature amplitude modulation) with greater bandwidth efficiency, use of software-intelligent coders and decoders, low production costs through the use of LSI technology as well as much lower power consumption of broadcasting and consumer equipment ('green products').

As a result of digital technology, professional equipment and consumer products will be more similar, so that for instance, the same LSI chips will be used in both categories of equipment.

But despite all these convincing arguments, digital technology would not have been successful without tackling two problems right from the start: the high data rate of the source signal at the output of the analogue/digital convertor and the problem of terrestrial multipath reception.

With for instance a sampling rate of 48 kHz and 16-bit linear coding, a stereo signal has a source data rate of about 1.5 Mbit/s. Assuming a baseband width of 45 MHz, threefold sampling frequency and a resolution of 8 bits, a HDTV signal features a data rate of about 1 Gbit/s. The direct transmission of such data rates for audio and video would not be viable economically . To solve these problems, new ways are being followed in R&D at the expense of perfectionism: at the destination (receiver) the signal is not reproduced, but the human ear and eye are 'fooled', using results from the science psychoacoustics and visual phenomena. The redundant and irrelevant signal components are not transmitted. In this way, audio data can be reduced by a factor of eight without affecting the CD quality and HDTV data by a factor of seven (reduction to 140 Mbit/s), or a factor of 30 (34 Mbit/s) up to a factor of 50 (20 Mbit/s). Although for audio data reduction with MUSICAM (masking-pattern-adapted universal subband integrated coding and multiplexing) a worldwide accepted standard has been established, the HDTV data reduction still covers a wide range: from 140 Mbit/s to approximately 20 Mbit/s. For the implementation of audio and video coders and decoders, DSP in conjunction with LSI plays an essential role.

The reason that digital technology for sound broadcasting with DAB (digital audio broadcasting) has not been implemented until now, is that until the mid-1980s experts did not see an economical solution to the problem of terrestrial multipath reception. It was at that time that a whole series of measures was first considered: multicarrier methods, an orthogonal carrier arrangement, a guard interval, time and frequency interleaving and a single frequency network (SFN). The DAB modulation method is known by the acronym OFDM (orthogonal frequency division and multiplexing). In field trials the multipath channel (Rayleigh channel) was thus finally mastered with echo signals being used as wanted signals. This method was first used in DAB. The step from DAB to DIB (digital integrated broadcasting) for instance, where audio, video and value-added data, such as radio data systems, RDS, radio paging, RP, systems or traffic information data (traffic message channel, TMC) are rated as anonymous data for a transparent data channel, is a simple one.

The necessity of mobile reception of sound programme signals in concert quality, with great operating convenience and value-added services is undisputed in the world of technology. But what about the TV signal, with the driver of a car not being able to watch television? There are already applications for mobile TV reception, as for instance in buses, caravans, trains, VIP cars, taxis or mobile offices. With the aid of data reduction, various quality levels from HDTV quality (34 to 20 Mbit/s), PAL/D2-MAC quality (10 Mbit/s) down to watchman quality (the video equivalent of the walkman) with approximately 2 Mbit/s and the quality of the video telephone (64 kbit/s) can be implemented.

In fact, in the future, television can be expected to provide cinema quality pictures using HDTV, stronger impact gained from large picture projects, readily labelled 'telepresence', will be possible and TV developments will range from greatly improved TV pictures to the introduction of low quality TV for mobile reception with small screens. The chief elements of the sound and TV broadcasting revolution are shown in Table I.2.

Components	Analogue	Digital
Amplifier technology	Vacuum	Solid State
Technology S-curve		
Modulation	AM, FM	PSK, QAM
Signal	Reproduction	Impression
Channels	$CH_n \cdots CH_m$	Single frequency network

Table I.2 Elements of the sound broadcasting and television reform

Acknowledgements

The author wishes to express his gratitude to Karin Bruckmoser, Claudia Fritz and Corbinia Dambacher for their assistance in text and graphics processing (Winword 2.0, ME10/UNIX) as well as Dr. Jürgen Lauterjung and Albert Winter for their helpful and encouraging discussions.

Chapter 1

Sound and TV broadcasting media

The term media in connection with sound and TV broadcasting is widely understood as the 'new media', created by private programme providers, which previously could only be received via satellite or cable. Now, sound and TV programmes can be distributed via four media: terrestrial transmitter networks, direct broadcasting satellites and line-conducted signals via copper coaxial and fibre optic cables.

The terrestrial and uniform coverage by public broadcasters in Europe(such as ARD and ZDF in Germany, RAI in Italy, the BBC in the UK and TDF in France) was largely completed in the mid-1980s. At this time the expansion of basic network transmitters (with powers of typically 5 kW to 20 kW) came more or less to a halt and further investments were made almost exclusively to fill gaps in the coverage (transmitter powers of less than 200 W). It was only when the private programme providers received approval for terrestrial broadcasting, from 1988 in Germany for example, that further transmitters were installed. The greatest problem for the national frequency allocation authorities was to find free channels [19].

It is intended in the near future to release channels 61 to 69, which were reserved for NATO, for public utilisation by broadcasting services. National and international committees are consulting about the use of these channels for TV programmes. It seems that these frequencies will be reserved for a future TV standard (digital HDTV).

Reception from direct broadcasting satellites (DBSs) is today in open competition with cable and terrestrial channels, and under the current laws the consumer is left with a free choice. The breakthrough to DBS came with the medium-power satellite ASTRA, which allows reception with a dish antenna of as little as 60 cm in diameter and provides 16 channels. ASTRA has gained a market share of 70% in DBS alone and with combined receiving systems ASTRA/Kopernikus of more than 90%.

The development of broadband cable networks has been advancing in Europe at different rates, the number of households connected lagging behind the real capacity of the networks. Today, the subscriber density (ratio of number of households connected to maximum number of connections possible) is for instance approximately 55% in Germany. This low density and the connection fees that have been calculated too low have resulted in an undercoverage which in turn leads to reductions in investment. A greater number of subscribers could only be achieved if the broadband communication network provided more free channels (hyperband and lower UHF range). This would give an impetus to new, local programme providers to appear in the marketplace.

In the dual system of sound and television broadcasting a change to favour cable and satellite as well as private programme providers can be seen throughout Europe. This is shown by its acceptance expressed in terms of viewing figures and the directly related advertising revenues. Despite this development, terrestrial transmitter networks for sound and television broadcasting will keep their key position in the long term because of their special performance features. They provide basic coverage to the local and regional service areas, shadowed areas and — in the longer term — coverage for the mobile reception of data, music and moving pictures. Fig. 1.1 gives an overview of the technical and technological advances of the individual types of media.

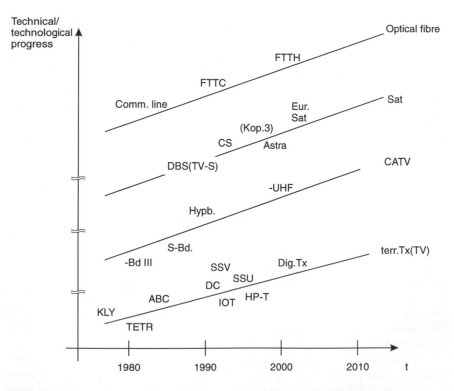

Fig. 1.1 The technical media for sound and television broadcasting

terr. Tx - terrestrial transmitter
KLY - Klystron
TETR - Tetrode
ABC - Annular beam control
HP-T - High power tetrode
MSDC - Multi stage depressed collector
SSV - Solid state VHF
SSU - Solid state UHF
IOT - Inductive output tube
Sat - Satellite
DBS - Direct broadcasting satellite
    (e.g. TV Sat)
CS   - Communication satellite
    (e.g. Kopernikus)
Eur.s. - European satellite

CATV - Cable authority TV
-BdIII - including Band III
S-Bd - special channel
Hypb. - hyperband
-UHF - including UHF

Fibre - Optical fibre
Contr - Contribution
FTTC - Fibre to the curb
FTTH - Fibre to the home

1.1 Terrestrial transmitters

1.1.1 VHF FM transmitters

The most important function of the VHF FM transmitters today is the provision of coverage to mobile receivers. This has to do with the increased importance of the car in society as a whole and applies even though VHF sound broadcasting was meant for stationary reception with an antenna height of 10 m. It is therefore inevitable that VHF sound broadcasting no longer satisfies today's requirements regarding quality and ease of operation, especially since the compact disc has set such a high standard in this field.

This need prepared the ground for DAB which will, however, not be introduced until after 1995 and simulcast operation at VHF is being planned. Some in the United States see DAB as a supplement to the established FM sound broadcasting. CD radio, a term used in the US, will not replace FM in America which does not currently dominate AM sound broadcasting. As such VHF sound broadcasting will probably remain in operation for at least another 20 years. Consequently, the associated technologies (for example amplifier technology and radio data system, RDS) will continue to exist. The continuation of VHF sound broadcasting is of significance for DAB too. Although DAB requires highly linear amplifiers instead of the Class-C switching amplifiers used in FM, and the modulation method changes from analogue frequency modulation to PSK and multicarrier methods, there is nevertheless a continuity in amplifier technology from FM transmitters to DAB transmitters. With FM transmitters in the 10 kW category, the trend is towards solid-state amplifiers as an alternative to tetrode amplifiers. Solid-state amplifiers are also required in DAB although here they are used at a lower average output power (<1 kW). Due to the multicarrier method used in DAB, a modulation signal similar to white noise is obtained, so the amplifiers must feature an extremely wide dynamic range. They are therefore designed for a crest factor (peak power to average power) of 5 to 10. In the following a 10 kW tetrode transmitter and a 10 kW solid-state transmitter are taken as an example to describe the development of the VHF FM transmitter technology.

1.1.1.1 10 kW VHF FM tetrode transmitter

The development of broadcasting transmitter systems began with the introduction of VHF sound broadcasting in the Federal Republic of Germany in 1949. The first 10 kW transmitter came on the market early in the 1950s. The output stage was made up of a triode. Power tetrodes later made it possible to accommodate a 10 kW transmitter in a single rack. The main features of a modern tetrode transmitter are [2]:

- Exciter with synthesiser and built-in stereocoder
- High efficiency
- Electronic filament power regulation
- Power tube with long lifetime
- Optional use of at least two types of tubes (second source)
- Internal or external ventilation producing low noise levels
- Continuously tunable in VHF FM range requiring two tuning elements only.

The quality determining element of the transmitter is the exciter. It contains a remotely controllable synthesiser oscillator and is suitable for use with transmitter systems in active, passive or (n+1) standby configuration. The output frequency can be set in 10 kHz steps between 87.5 and 108 MHz. In addition to mono and stereo signals, additional information such as traffic programme or SCA (subsidiary channel authorisation) signals and RDS signals can be transmitted [7].

The 10 kW amplifier of the transmitter is made up of the following modules: a driver amplifier, a 10 kW cavity resonator and a harmonic filter (Fig. 1.2). A push-pull amplifier with field-effect transistors that produces the necessary driving power of approximately 50 W in a single stage may be used as the driver stage. The 10 kW cavity resonator contains the cathode circuit and the anode circuit. The cathode circuit is of broadband design so that only two elements have to be tuned to change the frequency: an inductive anode circuit and an inductive output coupling (Fig. 1.2). These tuning elements are two identical shorting plungers allowing the inductances of a stripline to be continuously varied over the entire frequency range.

Tube replacement should be fast and easy. By appropriate installation of the coaxial, three-section harmonic filter, the RF output can be brought out at the top or bottom of the rack. For the 10 kW amplifier, a circuit with grounded centre tap of the coil between grid and cathode should preferably be used. It combines the advantages of a grounded grid circuit with its high stability and a grounded cathode circuit with its low driving power requirements. In addition, this circuit can easily be neutralised by balancing the interelectrode capacitances to form a bridge circuit.

The new output tube RS2068CL developed by Siemens ensures a long lifetime and large power reserves as compared with previous tetrodes. The following measures have been taken to enhance the average life (Fig. 1.3):

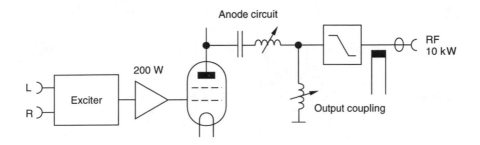

Fig. 1.2 Basic circuit diagram of 10 kW VHF FM tube transmitter

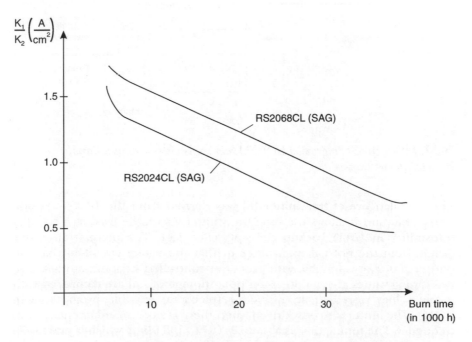

Fig. 1.3 Cathode current relative to cathode surface (K_1/K_2) as a function of burning hours at constant filament power (lifetime of cathode emission) [2]

- Enlargement of the cathode surface, which reduces the emission current density by 25%
- Reduction of the cathode temperature by low peak operating current
- Change in technology of the grid coverage to molybdenum zirconium-carbide platinum. The mechanical and thermal stability of the electrode could thus be improved to avoid geometric spreading
- Improvement of air flow for cooling metal ceramic compounds.

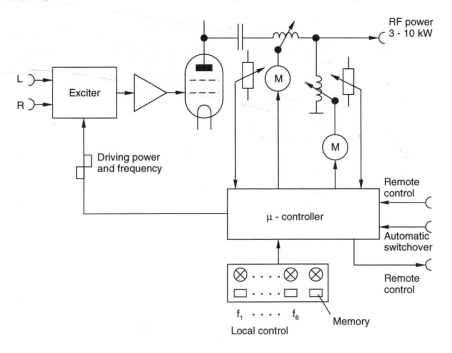

Fig. 1.4 Basic circuit diagram of UHF-FM tube transmitter used as a standby in (n+1) concepts (preset transmitter)

The so-called **preset transmitter** [3] was derived from the 10 kW tetrode transmitter and it allows the tube transmitter also to be used as a standby transmitter in (n+1) backup concepts (Fig. 1.4). The preset transmitter differs from the normal transmitter in that the anode circuit in that the output coupling is tunable with processor-controlled stepping motors. The two tuning values are two precision potentiometers and are compared with preset values stored in an EEPROM (electrical erasable programmable ROM). The motors are only driven when there is a command to change the frequency. The tuning time is about 10s (87 to 108 MHz) which is practically the same as the warmup time of the standby transmitter. Once the standby has taken over operation from the main transmitter, this preset transmitter acts like a conventional tube transmitter. Every local tuning of the transmitter on the antenna can be transferred to the EEPROM and updated at any time. A frequency can be changed by remote control, local control or by favoured automatic switchover. The switchover can work with six main transmitters of different frequency and output power (3 to 10 kW). Even without automatic switchover, the transmitter can operate independently and with selectable frequency.

One of the advantages of this concept is that tube transmitters can be manufactured more cheaply than broadband, solid-state transmitters which were commonly used as (n+1) standby transmitters in the past. The previous standby concept used for tube transmitters, the passive standby, can thus be disregarded. In the passive standby system, an automatic switchover unit switches to a second, identical transmitter whenever the main transmitter is out of service.

1.1.1.2 10 kW VHF FM solid-state transmitter

The great leap forward to low-priced transistorised transmitters of the upper power class has become possible thanks to new MOSFET transistors, for instance BLF278 from Philips-Valvo or MRF1516 from Motorola, with a power gain of about 20 dB on a supply voltage of 40 to 50 V. Previously transistors were based on the NPN bipolar technology and featured a gain of 8 to 9 dB in Class-C push-pull mode (for instance TRW-Motorola Type TP9383) [6]. The use of MOSFET transistors in conjunction with state-of-the-art PCB manufacturing methods allows RF amplifier boards with printed RF power components such as capacitors, inductors and lines as well as fixed power splitters and couplers to be produced.

The operating principle of the solid-state transmitter is described below using a 10 kW transmitter (Fig. 1.5) [5] as an example: the driver power from the 20 W exciter is routed to the eight amplifier modules via Wilkinson dividers. The high efficiency plus optimised compressed-air cooling of the amplifier modules permit very small heat sinks to be used. Identical couplers combine four amplifiers to yield an RF power of 5 kW. The two 5 kW outputs are combined via the 3 dB power coupler to give 10 kW [4]. After disconnecting its driver power, an amplifier can be withdrawn while the transmitter is still in operation.

A microprocessor-based transmitter control unit provides the correct switch-on sequence, monitoring, display and LED indications and remote control of the transmitter. For servicing, it is possible to switch the transmitter control unit to manual mode, meaning that the switch-on sequence 'fan on' through to 'carrier enable' is performed manually. Thanks to the modular design of the 10 kW transmitter, configurations with 2 × 5 kW single transmitters, 2 × 5 kW dual transmitters (passive standby) and 2 × 5 kW single transmitters with (n+1) standby can be implemented [5].

In the following, the **1.3 kW amplifier module** is examined in greater detail. This rackmount module comprises a regulated MOSFET driver stage, a 1:6 power divider, six 250 W push-pull amplifiers, a 6:1 power coupler, a

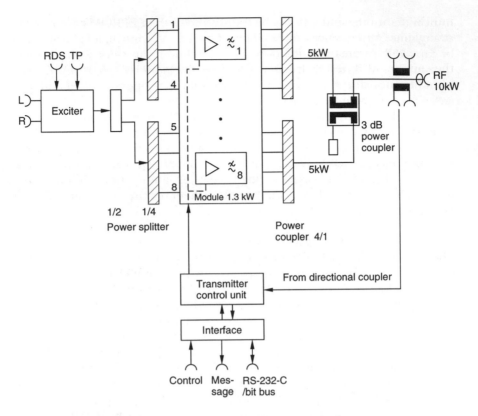

Fig. 1.5 Block diagram of 10 kW VHF FM solid-state transmitter

harmonic filter and a monitor. The 250 W push-pull amplifiers are fitted with a 300 W dual MOSFET; this ensures a safe margin with respect to the transistor ratings under demanding operating conditions such as a VSWR (voltage standing wave ratio) ≥ 1.5 and an inlet air temperature up to 50°C. The power splitters and couplers are printed on Teflon boards and require no alignment. The harmonic filter provided in each amplifier plug-in also consists of a Teflon board with the printed power capacitors and a soldered series/parallel inductor hybrid (a meander cut from a copper plate with a laser tool).

The monitor implemented in SMD (surface mounted device) technology protects the amplifier from overloading and determines the operating status of the module to the transmitter control unit. If one of the power transistors fails, the monitor prevents undue power increase in the remaining 250 W power stages. For servicing, replacing the 1.3 kW module or the faulty 250 W MOSFET module (with transistor) or merely the defective transistor requires only the adjustment of the pinch-off voltage. Transistor failure is indicated on the display of the transmitter control unit and each amplifier module has an integrated lightning protection.

In a solid-state transmitter not only the amplifier is important, but also the coupling technique used. The **Wilkinson coupler** (Fig.1. 6) is particularly suited for printed circuits [8]. It is also called a 0°-coupler, since the two RF outputs to be combined are applied to the coupler in-phase.

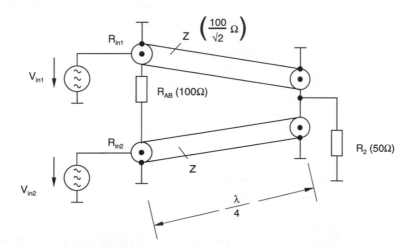

Fig. 1.6 Principle of Wilkinson coupler

For better understanding of the coupler, the load resistor R_L is assumed to be a parallel circuit of two 100 Ω resistors which are transformed to 50 Ω at the inputs via the $\lambda/4$ line with Z = $100/\sqrt{2}$ Ω. No power is absorbed by the resistor R_{AB}, since the voltages at its two terminals are equal. The disadvantage with this type of coupler is that the dummy resistors which in the case of a fault (magnitude and/or phase error) absorb the differential power of the fed-in RF power are connected at high-end voltage between the ports. To implement the unbalanced and, hence, the connection of the absorber resistor anywhere with optimal cooling, a three-port with bandpass characteristics is used within the operating frequency range [9].

1.1.2 TV transmitters

The terrestrial TV transmitter technology in Europe went hand in hand with the development of television. For more than 20 years this was the only medium of growth. Over the past 10 years the terrestrial TV transmitter network has been in competition with the new broadband communication media: coaxial copper cable and broadcasting satellite. Similarly, as with cable and satellite reception, TV transmitters will probably remain important over the next decade for TV sets designed for stationary reception. This class of equipment includes portables, which — although being more flexible regarding their place of use — still require alignment of the associated aerial and a power outlet.

The handheld pocket TV is offered to an increasing extent on the market at a price below £400. The development of these 'watchman' units shows that the consumer industry sees a market for quasi-mobile reception although reception from terrestrial transmitters has been specifically made for stationary equipment.

Future digital TV transmitter networks will, however, have a different task. They must be suitable for portable and — with limited picture quality — for mobile reception. That is precisely why the terrestrial TV transmitter technology will keep its place in the market. The digital TV transmitter networks will employ multicarrier methods requiring highly linear output stages with high overdrive capacity. For this reason and also to achieve economical, 'green' products that are aimed at, the TV transmitter technology will continue.

Transmitters using different technologies and their possible future developments are described below:

1.1.2.1 20 kW UHF TV transmitter with klystron

The third generation of 20 kW TV transmitters with klystron was introduced in Germany in 1983. These transmitters are used to transmit vision and sound signals in the frequency range 470 to 860 MHz in basic TV networks. The dual-transmitter system with a central control unit is for passive standby operation with automatic switchover in case of failure. The special features of a modern klystron transmitter include:

- Solid-state exciter stages and broadband amplifiers
- Multichannel sound operation
- Stabilised output power of vision and sound
- Precorrection of all nonlinearities at the IF
- Low-loss TV diplexer
- Dual-circuit vapour condensation cooling
- Utilisation of dissipated heat of klystron for heating buildings or warm water supply.

The single transmitter consists mainly of the vision/sound exciter stages, the broadband amplifiers, the power amplifiers for vision and sound, the TV diplexer, the power supply and the transmitter control units (Fig. 1.7) [11].

The vision/sound exciter stages include the modulation and deliver the output signals at the channel frequency. Modulation takes place at the standard intermediate frequencies: 38.9 MHz for the vision signal, 33.4 MHz for sound signal 1 and 33.158 MHz for sound signal 2. Nonlinearities of the signal path are corrected at the IF. The crystal-stabilised local oscillator is

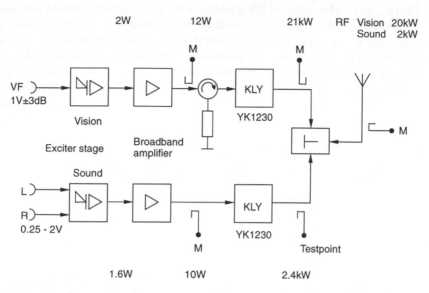

Fig. 1.7 Block diagram of 20 kW klystron transmitter

used both for the vision and for the sound section and produces the required frequency for converting the IF to the channel frequency.

Each exciter stage is followed by a 15 W broadband transistor amplifier which drives the power amplifiers. Four-cavity klystrons YK 1230 with electromagnetic focusing are used for the vision and sound power amplification. The output powers are kept constant by stabilising circuits. Black-level control is used in the vision section and the TV diplexer combines the vision and sound output powers.

The dissipated power on the collectors of the two klystrons is eliminated by an almost noiseless dual-circuit vapour condensation cooling system. A heat exchanger transfers the heat to a secondary circuit with nonfreezing industrial water, from where it is dissipated to the atmosphere. A second heat exchanger in the secondary circuit allows the dissipated heat to be used for heating the building or hot water supply [13].

1.1.2.2 Klystron with ABC

ABC (annular beam control) increases the efficiency of klystron transmitters using the newly developed klystrons (e.g. YK 1233) [10]: these klystrons have an annular electrode near the electron gun permitting control of the beam current with a few hundred volts. In a similar way to the AB operation of power tetrodes, the ABC mode permits the maximum RF power to be produced during the sync. pulse and the beam current determining the power consumption to be reduced during the active picture content. Using an ABC klystron, the power consumption of a 20 kW UHF TV transmitter, for instance, can be reduced from 78 kVA to approximately 68 kVA. The basic circuit diagram of the ABC control is shown in Fig. 1.8 [14]. In the VF amplifier, the sync. pulse required for the ABC delay controller is derived from the input sync. pulse. The ABC delay controller produces from this pulse the control pulses with different delay for the level control in the vision modulator, the phase modulator in the vestigial sideband filter and the ABC pulse unit.

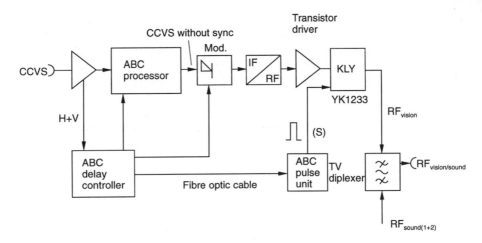

Fig. 1.8 Principle of TV transmitter with ABC unit

The ABC processor contains the black-level limiter, the video corrector and the sync. pulse separator. The control pulse is transmitted to the ABC pulse unit in the output stage via a fibre optic cable, since the ABC pulse unit is at a high potential. The vision modulator unit contains a phase modulator for correction of the phase error that additionally occurs in the output stage, due to the ABC mode.

The modulated vision RF signal is taken to the exciter which, via the selective output filter, drives the subsequent solid-state driver amplifier. The latter supplies the drive power for the vision output stage which is fitted with a klystron YK 1233 providing an output power of 20 kW. The klystron has four cavity resonators with electromagnetic focusing. Electrons are emitted by the cathode and accelerated. With the aid of the ABC electrode around the cathode, the beam current can be reduced. It is then taken to the input resonator and velocity modulated. In the second resonator, the beam is further velocity modulated and consequently the beam electrons are further bunched. This process is repeated in the third resonator until a power gain of more than 35 dB is obtained. In the fourth resonator, the electron density reaches its maximum. A large part of the beam power is weakened by deceleration of the electron packets due to the counteracting of the resonator field, and the DC power is so converted into RF power. Finally, the beam is absorbed by the collector and the kinetic energy converted into heat to be absorbed by the cooling system. The bandwidth and gain required for a TV amplifier can be obtained by suitable tuning and damping of the four resonators.

The operation of a TV transmitter with ABC-controlled klystron has to be adapted for PALplus signals (Section 3.4.1). The vertical helper signals of PALplus attain the blacker-than-black level and affect the ABC delay control. Primarily, the black level limiting has to be shifted. Another possible solution is to shift the helper signals at the transmitter input to the grey level range of the picture and clamp them to the black level again in the ABC-controlled klystron [15].

1.1.2.3 20 kW UHF TV transmitter with tetrode

The advantages of the tetrode transmitter over the klystron transmitter are its small dimensions and low power consumption. On the other hand, the tetrode transmitter comprises a larger number of tuned stages, each with its own power supply. Special features of the tetrode transmitter are [12]:

- High efficiency (about 40% lower power consumption than the klystron transmitter)
- Little space requirement due to compact design (about one-third the size of a klystron transmitter)
- Modern tetrodes in pyrobloc technique used in the power stages
- Dual-circuit vapour condensation cooling
- Utilisation of dissipated heat from the tubes
- Stabilised output power of vision and sound
- High basic linearity in tube stages
- Precorrection of all nonlinearities at the IF
- The transmitter uses separate vision and sound amplification (Fig. 1.9) [13].

In the vision section, two modern tetrodes, TH 527 and TH 582, based on the pyrobloc technique, are used in coaxial-line resonators. These amplifiers feature high linearity so that precorrection at the IF is easy and extremely high stability of the vision transmitter is ensured. In the sound section, tetrode TH 893 is used in a coaxial-line resonator. Similarly, the high linearity of this amplifier is of great advantage in multichannel operation in line with the IRT (IRT: Institut für Rundfunktechnik – Institute for Broadcasting Technology) dual-carrier method since it allows simple precorrection and high stability. Vision and sound signals are combined in a bridge-type diplexer with low attenuation and mutual isolation. An external cooling system with pressure, extractor fan and air mixer is provided for rack cooling of the TV transmitter. The dissipated heat of the tubes is absorbed in a dual-circuit vapour condensation cooling system. The primary circuit with distilled water is connected to secondary circuit with a nonfreezing medium via a heat exchanger. Through a further heat exchanger in the secondary circuit, the dissipated heat can be utilised for heating buildings or hot water supply.

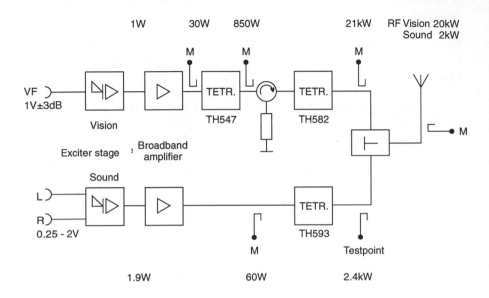

Fig. 1.9 Block diagram of 20 kW tetrode transmitter

1.1.2.4 10 kW VHF TV transmitter with transistors

This solid-state TV transmitter allows dual-sound and stereo-sound television broadcasting using the principle of separate vision and sound amplification in Band III: 170 to 230 MHz. The vision and sound signals from the power amplifiers are combined by a TV diplexer. Vision and sound amplifiers operate in active standby mode. A passive standby system (dual drive) as shown in Fig. 1.10 is usually adopted for the exciter stages. The power amplifiers use bipolar transistors (transistor SD1485 from SGS Thomson) which due to their higher stability compared with MOS transistors are more advantageous for linear amplifiers. The transistor power amplifiers have effective self-protection facilities responding in case of input overload, high reflections at the output and excessive temperatures.

The output power of the 10 kW vision amplifier (sync. peak) is produced by eight rackmounts connected in parallel. The sound amplifier comprises two modules. The power amplifier is made up of six modules: the predriver and the driver modules and four identical output amplifier modules. All modules are of broadband design for the frequency range 170 to 230 MHz

[16]. Each amplifier is powered by a power supply unit with primary switched-mode regulators to provide high efficiency. Size and weight are minimised by the high switching frequency, while the power consumption of the transmitter is about 30 kW.

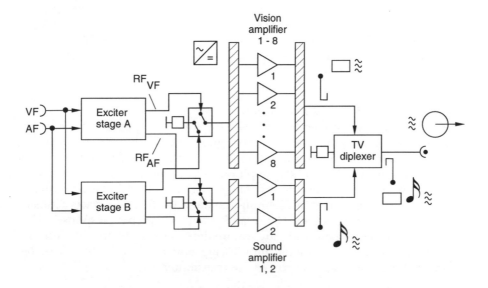

Fig. 1.10 Block diagram of 10 kW TV solid-state transmitter for Band III

For maintenance, the amplifier rackmounts and power supply units can easily be replaced while the transmitter is in operation. In case of a failure, the faulty module will automatically be indicated. On replacement of a module, the output power is briefly reduced and as no alignment is required, optimum on-air time is ensured.

The solid-state transmitter uses forced-air cooling, the power amplifiers being cooled by a high-pressure cooling system producing an air flow of approximately 80 m^3/min. The high-pressure system achieves a distinctly better efficiency than the pure low-pressure ventilation systems. The temperatures of the power transistors are thus kept permanently low. This results in a long service life and high reliability.

The following are typical requirements placed on the transmitter control and TV exciter:

Transmitter control:

- Multilingual menu guidance
- Graphic display
- Nonvolatile storage of all settings, fault messages and counts in plain text

- IEC/IEEE bus, RS-232-C and bitbus interfaces.

TV exciter:

- Fully electronic tuning
- Menu-guided settings
- Microprocessor control
- Synthesiser and precision offset capability
- Dual-sound capability in line with IRT method and NICAM 728
- Integrated stereocoder
- Sync. pulse regeneration.

1.1.2.5 10 kW solid-state UHF transmitter
This transmitter is suitable for all TV standards with negative modulation (highest RF carrier amplitude with sync. pulse) and multitone systems [17]. In the TV exciter, the video and audio signals are modulated to yield separate IF signals, converted into RF, amplified separately and combined in a diplexer to give the output power. The vestigial sideband characteristic is implemented at the IF by a niobium-titanate SAW filter (surface acoustic waveform) providing an excellent and stable frequency response. The transmitter is equipped with an automatic control system for use in unattended stations. Additional monitoring and control facilities can be connected and the main features of the transmitter are:

- Four types of amplifiers for the UHF band
- High-power RF transistor with 150 W output power (SD1492 from SGS Thomson)
- Module replacement without tuning or alignment while transmitter is in operation
- All RF and IF connections have an impedance of 50 Ω
- Built-in measurement, monitoring and remote control functions
- 10 kW model in 2 racks (H × W × D = 1.9 m × 1.6 m × 1.5 m)
- Power consumption for 10 kW TV power 32.4 kW for APL (average picture level)
- Standby concept: exciter standby (as shown in Fig. 1.11), active standby and passive standby
- Air cooling 5000 m^3/h.

1.1.2.6 Advances in TV transmitter technology
Table 1.1*a* shows the technology tree of the amplifier techniques used in terrestrial TV transmitters [20]. Solid-state transmitters are developed typically for a power class up to 10 kW (in special cases 20 kW). In Band III, MOS transistors (e.g. SK410 from Toshiba, BLF378 from Philips) or bipolar transistors (e.g. SD1485 from SGS Thomson) are used by transmitter manufacturers. In Bands IV/V, bipolar transistors (e.g. SD1492 from SGS Thomson, MRF1516 from Motorola) are preferable. Regarding vacuum

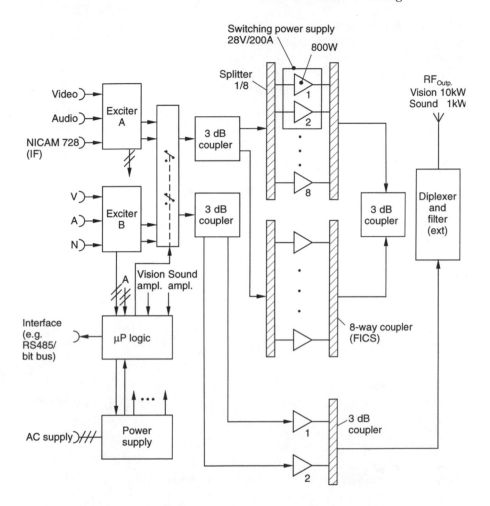

Fig. 1.11 Basic circuit diagram of 10 kW solid-state UHF TV transmitter [18] (FICS: fully isolated coupling system)

devices, a tetrode for the power from 40 to 60 kW is being further developed (Thomson).

The klystrode (Varian, US) combines the technology of a tetrode and a klystron and has been further developed to the IOT (EEV, England). The IOT (inductive output tube) [21] is a double-tuned tube whose output resonator is coupled to the collector circuit with an inductance. The IOT is an electron beam tube like the klystron and has a control grid like the tetrode. It was developed with the aim of combining the advantages of the basic technologies of a tetrode [22] with a klystron:

- Low production cost
- High efficiency (cost of ownership)
- Same dimensions as tetrode
- Easy to use
- Same lifetime and output power as klystron.

The klystron is also being further developed to obtain a higher efficiency. This is achieved by energy recuperation from the electron beam after the amplification process is completed. The method is based on several collector stages with increasing negative voltage, so the electron beam is decelerated.

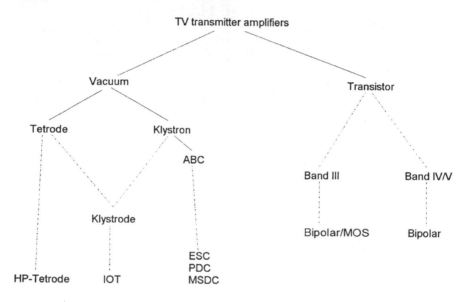

Table 1.1a. Technology tree for TV transmitter amplifiers

A further milestone in the refinement of the klystron is the collection of the electron beam at the collector so avoiding secondary electron emission which would reduce efficiency. The following designations are used by the various manufacturers:

MSDC   Multistaged depressed collector (Varian) [23]
ESC   Energy saving collector (EEV)
PDC   Philips depressed collector

A comparison of the various amplifier technologies is shown in Table 1.1*b*.

Of great interest is the question of the type of transmitter that will dominate in the future, taking into account the concept of combined or separate vision and sound amplification (TV diplexer), the standby concept (e.g. passive standby or (n+1) standby), the cooling concept (e.g. convection, forced-air cooling or liquid-cooling methods) also further criteria such as:

• Technical data
• Operating costs
• Price
• Service costs
• Replacement costs
• Complexity of maintenance.

The optimum concept can only be developed in co-operation with the operators (telecommunications services and broadcasting corporations), the makers of the technology (manufacturers of amplifier modules) and the providers of the complete broadcasting equipment (TV transmitters).

	Klystron ABC	Tetrode	HP Tetrode	Klystron DC	IOT	SS
Power	up to 65 kW	up to 30 kW	40 kW	40 to 60 kW	40 to 60 kW	up to 40 kW
Cooling method (preferably)	Water	Vapour condensation	Vapour condensation	Air/distilled water	Air/distilled water	Air/ convection
Efficiency for program signal (guiding data)	35%	50%	60%	65%	63%	30-40%
Gain	35-38 dB	13-15 dB	13-15 dB	34-38 dB	18-21 dB	7-8.5 dB
Collector voltage	22 kV	7.5 kV	9.0 kV	22 kV	30 kV	28 V
Average lifetime	30 Th	8 Th	10 Th	35 Th	25 Th	>10 years (depending on specs)

Table 1.1b Comparison of amplifier technologies used in TV transmitters for the UHF range

Note: The types of transmitters described in this section have already been implemented. The reason that only general characteristics were given was to outline the technology of a typical product on the market rather than one manufacturer's individual concept.

1.2 Satellites

Since sound and television broadcasting satellites have to be constantly accessible for the transmission of information, they are placed in a geostationary orbit. An orbit of this type is attained when the centrifugal force and the force of attraction between the satellite and Earth (Newton's law of gravitation) are equal, which occurs at an altitude of 35,786 km above the equator. There is only one orbit that is circular in the equatorial plane and in which the satellite completes its circuit of the earth in exactly 24 hours; relative to the Earth, the satellite appears stationary. Its speed in the 266,000 km long orbit is about 11,000 km/h [26]. The accurate position of the satellite can be given in degrees of longitude using the reference line (0°) of the international prime meridian line from the North Pole via Greenwich (UK) to the South Pole. The degrees of longitude east of the meridian line are defined as east longitudes and degrees west as west longitudes. They each cover a range from 0° to 180°. The satellite in the geostationary orbit has the function of an extremely high transmitter mast allowing information to be transmitted to Earth at ultra-high frequencies up to, for instance, 20 GHz. The terrestrial area is covered by a beam (footprints). For the coverage of Europe, WARC 77 (World Administrative Radio Conference in 1977 in Geneva passed the following resolutions (Fig. 1.12):

- Europe, the then Soviet Union and Africa belong to coverage region 1.
- Up to 16 satellites can be positioned with an orbit separation of 6° allowing satellite receivers to receive broadcasts from several satellites of one group with just one antenna.
- The TV-SAT for coverage of Germany will be positioned at 19° west longitude in the orbit. Choosing this position for the satellite has the advantage that the so-called Earth's shadow effect occurs for about one hour at 2 o'clock in the morning, so the break in the direct power supply by the solar arrays is outside the main broadcasting time.

1.2.1 Telecommunications satellites

For reasons of quality, insensitivity to nonlinear transmission and because of the relatively high frequency bandwidth of the transponder (e.g. 27 MHz), frequency modulation is chosen for the transmission of television signals via satellite [24]. This applies both to the transmission via telecommunications satellites for programme exchange between broadcasting corporations and to the supply of the cable headends in broadband communication networks. For the so-called broadcasting satellites designed for direct reception via a parabolic reflector in the home, frequency modulation is also used. Communications satellites were originally intended only for point-to-point communications between individual countries. Their field of application was enhanced later on to include programme feeding to cable headends and also to individual receiving systems.

INTELSAT (International Telecommunication Satellite Organization) has been operating communications satellites since 1965. Originally, the communications satellites were operated in the 4 GHz range (C-band) only, with dish antennas up to 30 m in diameter being required. Now satellites are broadcasting to an increasing extent in the 11 and 12 GHz range (Ku band), allowing dish diameters of less than 10m for distribution. In Europe, EUTELSAT (European Telecommunication Satellite Organization) manages the operation of ECS (European Communication Satellite) systems. The frequency of the telecommunications satellites is 10.95 and 11.7 GHz downlink. National satellite systems like TELECOM (France) and KOPERNIKUS (Germany) radiate in the range 12.5 to 12.75 GHz. The EUTELSAT satellites ECS-1 and ECS-2 cover Eastern and Western Europe and permit receiving antennas with a maximum diameter of 3m to be used for programme distribution and feeding. The differences between communications satellites for programme distribution and direct broadcasting satellites have become less distinct due to the development of low-noise input stages and the resulting smaller antenna diameters. Today, individual reception of TV programmes from communications satellites is left to the consumer's discretion.

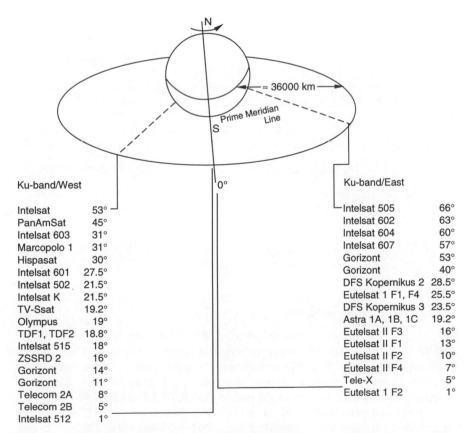

Ku-band/West		Ku-band/East	
Intelsat	53°	Intelsat 505	66°
PanAmSat	45°	Intelsat 602	63°
Intelsat 603	31°	Intelsat 604	60°
Marcopolo 1	31°	Intelsat 607	57°
Hispasat	30°	Gorizont	53°
Intelsat 601	27.5°	Gorizont	40°
Intelsat 502	21.5°	DFS Kopernikus 2	28.5°
Intelsat K	21.5°	Eutelsat 1 F1, F4	25.5°
TV-Ssat	19.2°	DFS Kopernikus 3	23.5°
Olympus	19°	Astra 1A, 1B, 1C	19.2°
TDF1, TDF2	18.8°	Eutelsat II F3	16°
Intelsat 515	18°	Eutelsat II F1	13°
ZSSRD 2	16°	Eutelsat II F2	10°
Gorizont	14°	Eutelsat II F4	7°
Gorizont	11°	Tele-X	5°
Telecom 2A	8°	Eutelsat 1 F2	1°
Telecom 2B	5°		
Intelsat 512	1°		

Fig. 1.12 Excerpt from satellite positions in geostationary orbit [28]

1.2.2 Direct broadcasting satellites

The DBS system has been designed for direct reception in a specific country or even over the whole of Europe [25]. At the WARC 77 conference it was decided to divide the frequency range 11.7 to 12.5 GHz into 40 channels with bandwidths of 27 MHz. There are two series of channels: the odd numbered channels with right-hand circular polarisation and the even numbered channels with left-hand circular polarisation (Fig. 1.13).

There is an overlap of about 19 MHz between the channels with right-hand and circular polarisation and those with left-hand circular polarisation. The channel allocation applies to various satellite positions, the position 19° west, for instance, has five channels for Germany, France, Austria, Luxembourg, Belgium, Netherlands, Switzerland and Italy. The transponders of the direct broadcasting satellites are rated for an output power of 200 W and initially permitted parabolic antennas of 90 cm in diameter. Today, input stages of higher sensitivity enable almost interference free TV reception with antenna diameters down to 40 cm.

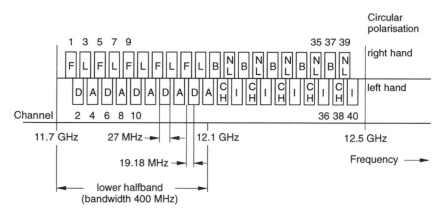

Fig. 1.13 *Direct broadcasting satellites in orbital position 19 ° west according to WARC 77 for direct reception in the individual countries*

1.2.3 Medium-power satellites

The Société Européenne des Satellites (SES) in Luxembourg introduced ASTRA, satellites with 16 transponders for direct reception. The ASTRA satellites allow 16 TV programmes and sound programmes on sound subcarriers to be received across the whole of Europe with parabolic antennas of less than 90 cm in diameter. ASTRA-1A has been in operation since 1989, -1B since 1991 and -1C since 1993. Satellites -1D, -1E and -1F for digital programmes are scheduled for 1994 to 1996 (see Table 1.2 and Section 4.8.3).

The individual satellites are positioned at a distance of several hundred kilometers from each other to minimise crossinterference to the orbital stability. The residual error can be corrected with the aid of orbit and position control systems. The enormous advantage of this closely positioned group of satellites is that it permits reception of several satellites with one reflector of fixed position. The users renting the ASTRA/SES channels (service providers) are private programme providers and to an increasing extent public broadcasting corporations. Due to the great number of programmes provided and the low cost of the receiving system, the ASTRA group constitutes the most attractive TV satellites.

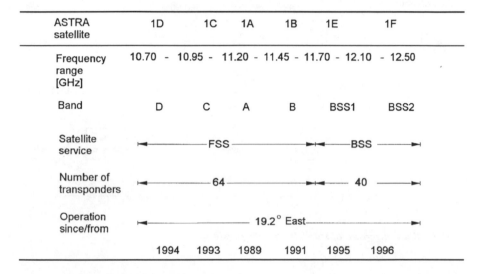

ASTRA satellite	1D	1C	1A	1B	1E	1F
Frequency range [GHz]	10.70 - 10.95 - 11.20 - 11.45 - 11.70 - 12.10 - 12.50					
Band	D	C	A	B	BSS1	BSS2
Satellite service	⊢————FSS————————⊢⊣—BSS———⊢					
Number of transponders	⊢————64————————⊢⊣—— 40 ——⊢					
Operation since/from	⊢——————— 19.2° East———————————⊢					
	1994	1993	1989	1991	1995	1996

Table 1.2 ASTRA system concept

The DFS (German communications satellite) system KOPERNIKUS, launched in 1989, is also a medium-power satellite operated at 12.5 to 12.45 GHz downlink. The orbit position of DFS2 is 28.5° east and that of DFS3 23.5°east. Discussions are being conducted at present to choose a power range for future di`rect reception satellites which falls between that of present-day broadcasting satellites with a maximum EIRP (equivalent isotropic radiated power) of 65 dBW and the EIRP of medium-power satellites of about 56 dBW, the so-called optimum power range [27].

1.2.4 Satellite receivers

The outdoor unit of a satellite receiving system includes the parabolic antenna with the orthogonal mode transducer and a mixer stage converting the received frequency band, 11.7 to 12.5 GHz with DBS, for instance, firstly into an IF of 950 to 1750 MHz (Fig. 1.14) [24]. In the indoor unit, a further mixer stage converts to the second IF, for instance, 480 MHz. At this frequency channel selection and frequency demodulation takes place

Following a European directive, the German broadcasting satellite TV-SAT2 and the French satellite TDF1, the same type, are using D2-MAC transmission MAC = multiplex analogue component) instead of PAL or SECAM transmission.

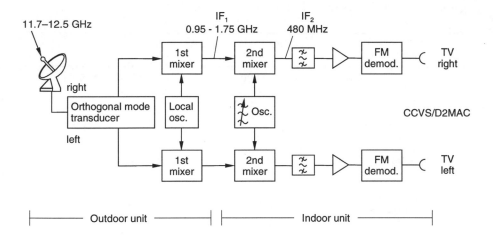

Fig. 14 Block diagram of satellite receiving system

1.3 CATV

The acronym CATV is not derived from cable TV, as is often thought, but means Community Authority TV. In 1948, the radio equipment dealer John Walson in the US installed his antenna at a distance of approximately 200 m from his receiver because of poor roof antenna reception and had to obtain permission for this installation from the local authorities.

In addition to radio links used as transmission media, cables have been used increasingly over the last ten years in Europe. Coaxial cables allow an almost interference free transmission of radio and TV programmes and also more programmes are offered via broadband cables than via terrestrial radio links. Although the initial investment costs for cable networks are much higher than for terrestrial transmitters, cabling increased during the early 1980s in the US cities, the Netherlands, Belgium and Switzerland, reaching a subscriber density of almost 90%. In Germany too, cable distribution systems were using these standard coaxial copper cables.

1.3.1 Cable headend

The programme is transmitted to the cable headends, which feed the cable networks, via satellites, terrestrial transmitter networks, microwave links or lines. National public broadcasting corporation programmes and private European and nonEuropean broadcaster programmes are fed into the cable networks. For the reception of ECS satellite signals for high-quality satellite receiving systems and for cable, TV sets for modulation conversion are required. The low transmitter power of the communications satellites requires highly complex receiving stations but these are justified only for professional broadband communications systems. Fig. 1.15 gives an overview of the various possibilities of distributing the sound and TV broadcast signals [29].

Satellite reception at the cable headend (Fig. 1.16) starts with the parabolic reflector, 3.7 m in diameter, for example. The frequency-modulated satellite signal at 11 GHz is converted into IF by convertors and taken via cable to the IF receiving system which demodulates the TV signal to obtain the vision and sound signals in the baseband. Direct conversion of the frequency-modulated satellite signals in the cable channels is not possible due to the required conversion from FM to AM. Therefore, the signals are modulated on new carrier frequencies in the range 47 to 300 MHz (PAL signals) before being distributed via the cable systems. To ensure compatibility with existing TV and sound broadcasting receivers, the TV programme signals are remodulated to vestigial sideband AM, commonly used in television, and the sound programme signals to FM, as used in VHF sound broadcasting. Special broadband communication TV and sound broadcasting transmitters are used for generating the channel signals, supplying a standard TV signal of 47 to 300 MHz. The transmitter contains vision and sound modulators in the standard IF range. In the output stage, the IF signals are converted to the cable channel. In the analogue dual-sound mode, the sound modulator produces the two carrier signals for stereo or dual-sound broadcasting. For a NICAM system, digital modulators are used with the appropriate selectivity characteristics to allow the interference-free operation of adjacent channels. The TV and sound broadcasting bands from the various sources are added up by a combiner.

1.3.2 Channel occupancy of broadband cable

The sound and TV broadcasting range of a broadband communication system is typically 47 to 300 MHz (Table 1.3) with the development of the D2-MAC standard making it necessary to introduce 12 MHz channels. These channels were planned to be above 300 MHz in the extended special channel range. This hyperband contains 12 channels, H21 to H32, with a 12 MHz bandwidth or 18 channels, S21 to S38, with an 8 MHz bandwidth allowing flexible occupancy. Presently consideration is being given to using the UHF

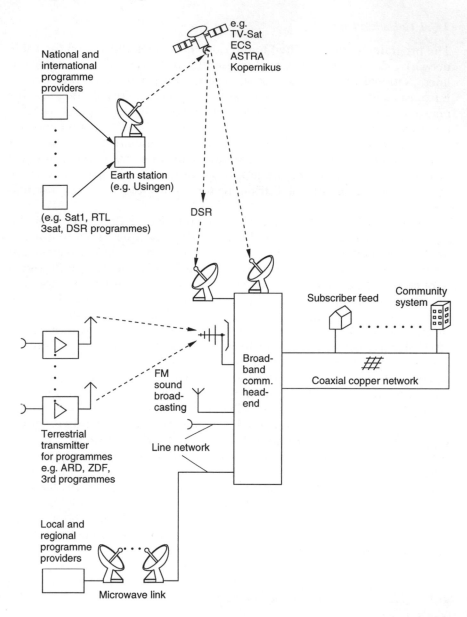

Fig. 1.15 Distribution of radio and TV programmes via the broadband communication network

Band, Bands IV/V, in the cable initially up to 600 MHz, and later in an even higher range. The much higher cable losses in this range have to be compensated for by amplifiers at correspondingly shorter distances. The extension of the cable frequencies will create additional possibilities for digital TV and future HDTV, and so cable has become a strategic option for the future.

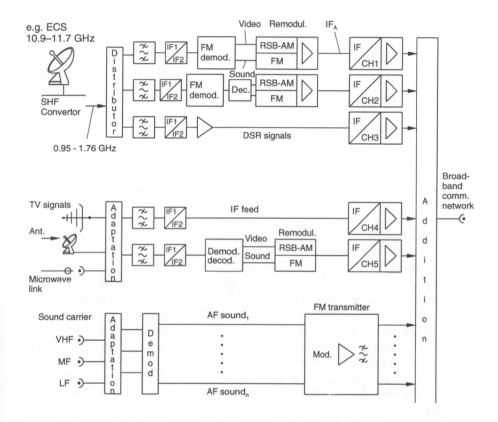

Fig. 1.16 Block diagram of a broadband communications headend

Band	Channels	Channels bandwidth MHz	Frequency range (MHz)	Signal	Range
I	2-4	7	47-88	PAL	VHF
II		7	87.5-108	30 stereo signals (VHF-FM)	VHF
Lower S-B	S2, S3	7	111-125	16 digital stereo signals (DSR)	VHF
Lower S-B	S4-S10	7	125-174	PAL	VHF
III	5-12	7	174-230	PAL	VHF
Upper S-B.	S11-S20	7	230-300	PAL	VHF
Hyperband	S21-S38	8	302-446	PAL (Dig.TV)	UHF
or	H21-H32	12	302-446	D2MAC, Dig. HDTV, DSR	UHF
[IV/V	33-60	8/12/24	446-790	(Dig. TV/Dig. HDTV)	UHF]

Table 1.3 Channel occupancy in CATV (S-B.: special channel band)

1.4 Fibre optics

In 1993 200,000 German homes were connected to fibre optic cable systems. This was the beginning of the commercial deployment following trials that took place in local telephone networks. The German DBP-Telekom was the first network operator to use fibre optic cable connections directly to the subscriber's home known as FTTH (fibre-to-the-home) or to the street distribution system known as FTTC (fibre-to-the-curb). The number of homes connected is now in the region of 1.2 million. The fibre optic cable transmits telephone and data services alongside cable TV programmes, integrating the previously separate telephone and broadband cable networks.

The technical standard of optical communications is high, considering that the attenuation of the light waves in optical fibres is less than 1 dB/km which is close to the theoretical limit. The development of suitable fibres has allowed transmission of single-mode signals which have low pulse dispersion so bit rates of 10 Gbit/s can be achieved with distances of more than 100 km possible without amplifiers. Laser diodes with small bandwidth and wide modulation range are being used and currently more than 35 TV signals can be transmitted on a single optic fibre.

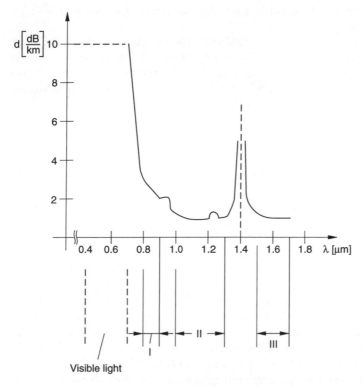

Fig. 1.17 Fibre attenuation as a function of wavelength [30]

1.4.1 Advantages of fibre optic cables

The main advantages of optical transmission are the low attenuation, the large bandwidth, the high immunity to interference and the high isolation from light signals of other fibres in the cable. This is the reason that today's long-range communications systems are using fibre optic rather than coaxial copper cables.

The intensity of a light beam in an optical fibre made of pure synthetic quartz glass is reduced by only about one-third over 1km [30]. Fig. 1.17 shows the ranges I, II and III in the near and medium infrared spectrum which are used for optical communications.

Compared with the frequency response of a coaxial cable used in cable TV networks, fibre optic cables have a wider bandwidth. Fibre optic cable networks now use single-mode fibres and laser diodes are the transmission source. Since the mode dispersion of the fibre is extremely low, the resulting bandwidth and length is more than 10 GHz × km. The disadvantage of fibre optic cables compared with copper cables, is that electro-optical and optoelectrical transducers have to be used at the beginning and end of the transmission link. These drawbacks are minimised by the large-scale production in microelectronics and digital transmissions increasingly becoming commonplace.

In 1989, the German company DBP-TELEKOM initiated the fibre optic switched broadband network VBN which allows the fibre optic overlay networks in about 30 cities to be linked by switched connections. This network features a bit rate of 139.264 Mbit/s and allows for instant digital transmission of video signals in HDTV quality.

1.4.2 Opto-technical components

The main components of optical fibre technology are: the fibre, the cable, the coupler and more recently the optical amplifier.

Today's fibre is exclusively standard single-moded, optimised for 1.3 µm [31]. This fibre is a high performance and broadband medium. The local area network quality of the fibres is less than 0.5 dB/km with a wavelength of 1.3 µm ±15 nm but they can also be used outside this wavelength. In the 1.3 µm range, bit rate × wavelength products of more than 200 Gbaud × km are possible depending on the fibre dispersion. Outside this wavelength range, the capacity drops to about one-fifth, but in absolute terms is still very high.

The main task of the **optical fibre cable** is to not only protect the fibres within it, but also to comply with the requirements of the local cable system such as fibre identification and distribution capability. A good cable design uses bundles of fibres in which, for instance, ten fibres coded by colour are bundled and enclosed in a plastic waveguide. These fibre bundles can be twisted together to permit modular cable configurations of up to 2000 fibres.

In the **optical fibre couplers** the optical power is evenly distributed to 2 to 16 outgoing fibres. Optical fibre couplers are used in typical PON (passive optical network) topologies. The optical fibre cable distributors are usually produced using a fibre melting process in which 2 to 8 fibres are laterally melted together. An alternative is the planar technique using a glass substrate with diffused optical fibre distributors [31]. In the longer term these planar couplers are expected to take over because of the advances in integrated optical technologies.

The **optical amplifier** is a recently developed optical fibre component. Optical amplifiers or fibre amplifiers are used extensively in high-speed experiments to cover greater distances or to achieve larger bit rate × wavelength products [32]. Through the use of optical amplifiers the bandwidth of the transmission link can be increased from the Gigahertz to the Terahertz range. The theoretically extremely large bandwidth of the optical fibre therefore proves more economical. For this purpose, the optical amplifier should have the following characteristics: transparency of data rate, modulation (digital and analogue) and the signal format NRZ (non return to zero), HDB3 (high density bipolar with maximum three successive zeros). The optical amplifier should also be suitable for wavelength multiplexed signals and insensitive to various light polarisation modes. Fibre optic amplifiers using erbium-doped single-mode fibres meet these requirements and are helping to bring about a revolution in optical transmission both in long-range traffic systems and subscriber lines [32]. Optical amplifiers allow cost the effective implementation of FTTC and FTTH systems, its operating principle is shown in Fig. 1.18. Alternatives to the optical amplifier are semiconductor amplifiers which have advantages regarding the noise, linearity and polarisation sensitivity parameters and further development can be expected for this type of amplifier.

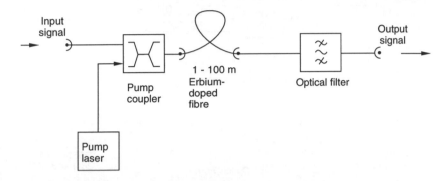

Fig. 1.18 Principle of optical amplifier [32]

1.4.3 Sound and TV broadcasting via fibre optic links

Currently, the transmission capacity of single-mode fibres is under used. Data rates reach 1 Gbit/s and more. Analogue signals are still used in VHF sound and TV broadcasting, because the linearity of the laser diodes has proved to be inadequate. Therefore, the signals have to be transmitted in digital form on fibre optic links using coders and decoders (Fig. 1.19). With SCM (single carrier multiplex) the light of one wavelength is modulated to 140 Mbit/s per TV programme. For 35 TV and 28 sound programmes at least nine fibres are needed. This is the minimum requirement for a broadband coverage offered by the German DBP-Telekom via broadband communications networks. A data reduction would result in a higher programme capacity, but with increased complexity through the use of codecs.

   An interesting method would be a transparent transmission of the signals via the optical fibre [33, 34], equally for digital and analogue modulation. For broadband, for instance, frequency-multiplexed signal with any modulation mode, the transmission link is subjected at the beginning to electro-optic conversion and at the end to optoelectric conversion. Optical intermediate amplifiers are a cost effective way of bringing the signal to the subscriber (Fig. 1.20).

In the US the aim of the long-term project 'Information Superhighway' is to link communities by an optical fibre network.

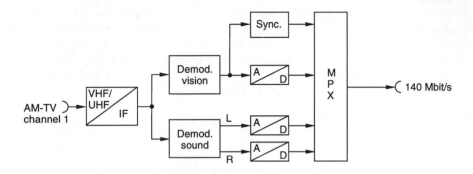

Fig. 1.19 Digital TV transmission via long lines

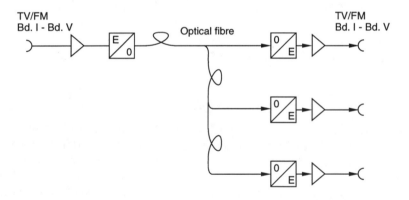

Fig. 1.20 Transparent transmission and distribution via optical fibres

Chapter 2

Technological advances in audio broadcasting

Terrestrial sound broadcasting using amplitude modulation in the long-, short- and medium-wave bands and frequency modulation in the VHF range has a long history. With the exception of stereo broadcasting for stationary VHF reception, the principle of sound transmission has remained more or less unchanged. The ARI (traffic information service for motorists) system introduced in Germany, Austria, Switzerland and neighbouring areas in France is a value-added service. RDS (radio data system), commencing in 1984 and introduced later throughout Europe with different applications of the information service, is a much more powerful system. With its introduction in the US in 1994 and with other countries outside Europe, RDS is about to become the worldwide standard (Fig. 2.1).

DSR (digital satellite radio) had already been specified at the end of the 1980s and its introduction planned accordingly. Due to technical problems — TV-SAT1 was unable to unfold its solar panels and was therefore unoperational — it was decided not to start the system across Europe as a whole, but only in Germany. From 1992 the powerful DSR method began to gain acceptance, and today its introduction in parts of Europe is providing a foundation for future technological advancement (Chapter 4).

The compact disc, launched in 1985 by Sony and Philips, was certainly one of the triggers of the digital revolution in sound broadcasting technology. The CD created a quality level that came very close to that of a live concert, and brought further advantages of easy and robust handling. The CD was not only suitable for the home but also for the car. The introduction of the CD challenged the current broadcasting technology to provide comparable quality and features. This led to the development of digital satellite radio for stationary reception via a parabolic antenna or coaxial copper cable network.

Finally, it was the CD which gave rise to the requirement for terrestrial use of digital audio for mobile reception. This requirement, which in the last decade was considered impossible because of the multipath problem in the Rayleigh channel, was suddenly fulfilled. The well known multicarrier method with an orthogonal carrier arrangement, causing low bit rates per carrier and hence long symbol periods relative to the propagation time, was used. This meant that the direct and the echo signals became compatible at the receiver.

Digital audio broadcasting (DAB) was specified between 1988 and 1992, with its introduction in Europe scheduled for the late 1990s.

DAB provides various other applications, such as traffic programmes, radio paging and transparent data transmission. A further development in this field is DB (data broadcasting), where open terrestrial broadcasting networks can be used for various audio and data applications.

In addition to the progress in satellite and terrestrial broadcasting, the CD has also led to further innovations in consumer audio equipment. These new products are described in detail in Sections 2.1.2–2.1.5, respectively: the DAT (digital audio tape), MOD (magneto-optical disc) recorder, DCC (digital compact cassette) and MD (mini disc).

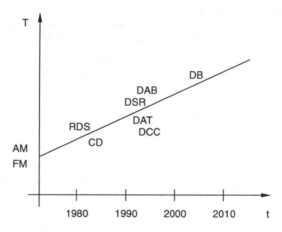

Fig. 2.1 The advance of audio and sound broadcasting (T = technology level)

2.1 Digital audio recorders

2.1.1 CD (compact disc)

Many revolutionary steps, such as digital signal storage, optical scanning, error correction and new manufacturing processes for both CD players and discs are embodied in CD technology. A CD contains digitally encoded music which is read by a laser beam. Since the laser is focused on a reflective layer embedded within the disc's substrate, dust and fingerprints on the surface will not affect reproduction. The effect of most errors that are likely to occur can be minimised by error correction circuits. Since no stylus touches the disc surface, there is no mechanical wear, no matter how often the disc is played. Thus digital storage, error protection and a long service life result in a robust hi-fi audio medium [37].

The history of CD began before 1974, when Philips studied the possibility of storing audio data on to an optical disc. Sony similarly explored the possibility of an audio disc and had extensively researched the error processing requirements. In 1979, Sony and Philips entered into a basic co-operation agreement and in June 1980 jointly proposed the Compact Disc Digital Audio System, with decisions on the signal format and disc material. Following the development of a semiconductor laser pickup and LSI circuits for signal processing and D/A conversion, the CD system was introduced in Japan and Europe in October 1982. In March 1983, the CD was made available in the US. By October 1983, the prices of CD players had dropped to about $400 (£250), with over 20 brands of players being available. In 1984, 900,000 CD players and 17 million CDs were sold, making the CD as the most successful consumer electronic product ever introduced into the market-place. Mobile and portable CD players have added to the range of products while CD players are now retailing around 200DM (£100).

The following CD system overview shows that the CD is a highly efficient information storage system (Fig. 2.2). The disc stores a stereo audio signal with 16 bit resolution and a sampling rate of 44.1 kHz. Thus 1.41 Mbit/s of audio data are processed by the CD player. Other processes such as error correction, synchronisation and modulation triple the number of bits stored on the disc. The resulting channel bit rate — the rate at which data are read from the disc — is 4.32 Mbit/s. A CD containing one hour of music holds 15.5 Gbits of information, of which 5 Gbits are audio data, on a disc of less than $4^3/_4$ inch (12.1 cm) in diameter [37].

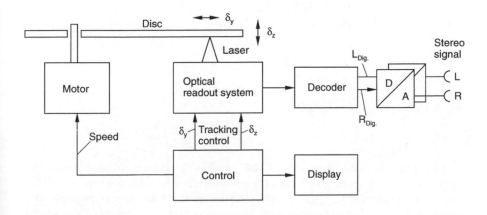

Fig. 2.2 Functional block diagram of CD player

Information is contained in the pits impressed into the plastic substrate of the CD (Fig. 2.3). This surface is coated with aluminium to reflect the laser beam reading the data from underneath the disc. A pit is about 0.5 µm wide and a disc may hold about 2 billion of them. The pits lie on a spiral track similar to the spiral groove in a conventional record. The pitch of the CD spiral, i.e. is how close together successive turns are, is, however, much greater than that on a conventional record: 60 CD pit tracks would fit into the width of one LP microgroove, which is about the width of a human hair. Each pit edge represents a binary 1; flat areas between pits or areas inside pits are decoded as binary 0. Data are read from the disc as changes in intensity of the reflected laser light.

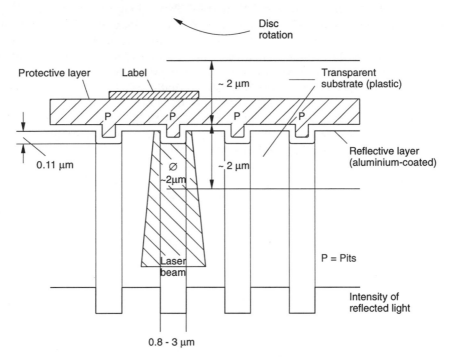

Fig. 2.3 Structure of the compact disc

The pits are encoded with EFM (eight-to-fourteen modulation) for greater storage density, and CIRC (cross-interleave Reed-Solomon code) for error correction [38]. Appropriate circuits in the CD players provide the demodulation and error correction. Because of the complexity of the signal processing, most players contain several microprocessor chips. When the audio data have been properly recovered from the disc and converted into binary signals, they are taken to D/A convertors and passed through the output filter and amplifier.

The CD system delivers hi-fi sound with outstanding technical specifications (Table 2.1). The frequency response is typically ±0.5 dB between 5 Hz and 20 kHz and the dynamic range is greater than 90 dB. The signal/noise ratio and channel separation at 1 kHz is greater than 90 dB. Figs. 2.3 and 2.4 show the structure and recording format of the CD.

CD-I (CD-interactive) [39] is a further development of the CD. CD-I is used as a storage medium similar to that of a CD-ROM, where video programmes on hobbies, management courses, educational topics, special interest and cultural subjects, entertainment etc. can be compressed and digitally stored. One CD-I can hold up to 650 Mbytes, providing 72 minutes video or 19 hours audio in AM quality. Stored in ASCII format (American standard code for information interchange), this capacity is adequate for 250 000 pages of text or 7 000 still frames; a portable version of CD-I is also available.

New opportunities are opening up for the world of colourful pictures and sounds, with the CD-I being heralded as the successor to the video disc.

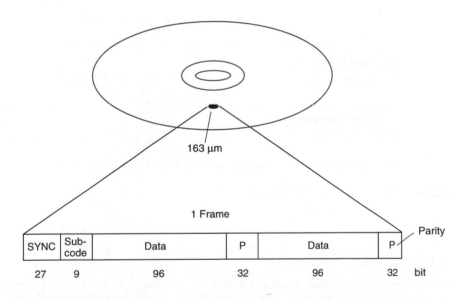

Fig. 2.4 CD recording format

A new aspect of CD development has become known as 'blue laser' [40]. Currently, disc players use red laser beams with a wavelength of approximately 780 nm. Sony has reduced the wavelength to approximately 470 nm, i.e. a blue beam, using a temperature of 200 degrees below zero and liquid nitrogen. If the blue laser were ready for application, the density of information held on a CD could be tripled. Sony predict, however, that it will take a few years for the blue laser to become suitable for use at room temperatures.

Disc	
Diameter	120 mm
Thickness	1.2 mm
Reading speed	1.2 to 1.4 m/s
Number of rotations	500 to 200 rpm
Playing time	max. 60 min (stereo)
Signal	
Sampling frequency	44.1 kHz
Quantisation	16 bit linear per L/R channel
Data rate from disc	4.3218 Mbit/s
Optical readout system	
Laser	semiconductor Al Ga As
Wavelength	790 nm
Audio performance	
Frequency range	20 Hz to 20 kHz; ±3 dB
S/N	>90 dB
Channel separation	>86 dB
Intermod. distortion	>86 dB
Phase linearity	±0.5°
THD	<0.05 %

Table 2.1 Technical data of CD system

2.1.2 DAT (digital audio tape)

As far back as 1985, manufacturers of consumer electronics were investigating digital magnetic tape recording methods. In addition to R-DAT (rotary head digital audio tape recording), S-Dat (S for stationary) was also examined. R-DAT was introduced, as with rotary heads an already existing technique could be used (Fig. 2.5). The R-DAT [42] offers:

• sampling frequency 48 kHz
• linear 16-bit quantisation
• playing time of 2 to 3 hours

- compact drive mechanism
- fast search
- smaller cassette than the conventional compact cassette
- additional data capacity allowing further applications.

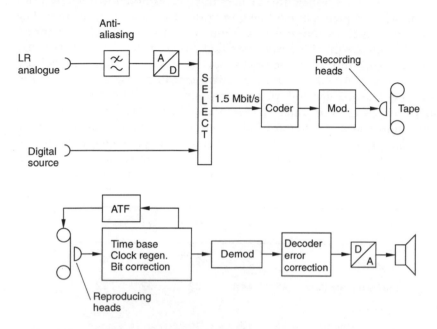

Fig. 2.5 Block diagram of R-DAT recorder

The R-DAT system uses a helical recording method. With an absolute tape speed of 8.15 mm/s, the R-DAT version requires only 60 m of tape for two hours playing time. The helical track written with PCM data has a width of 13.6 μm, which is about one tenth of the width of a human hair, and a length of 23.5 mm, with one data bit taking up 0.67 μm. The resulting data density of 17 Mbit/cm^2 could not have been attained with previous methods. Without reduction, the digital audio signals yield a data stream of about 1.5 Mbit/s. With error correction bits and various subcodes, this data stream is increased to 2.77 Mbit/s. For magnetic tape recording, error protection measures are taken in the form of ECC (error correction code) to avoid burst and random errors. Burst errors may be caused by coating defects on the tape, dust, scratches or contaminated heads. Random errors result from crosstalk from adjacent tracks, inadequately erased data or instabilities in tape transport. For R-DAT a double erasure correction Reed-Solomon code is used for error protection. This consists of an inner data code comprising data bits and parity bits, and an outer code surrounding the inner block with parity bit assignment in a fixed algorithm. The data blocks are additionally provided with interleaving and distributed to two adjacent tracks. Sync. bits, identification bits and the block address complement the data blocks.

Helical recording methods do not need separate control tracks for position and time reference. R-DAT uses the ATF (automatic track finding) principle with frequency bursts being recorded for track identification. The ATF system compares the crosstalk frequencies using analogue methods. The ATF zones are provided twice per track to avoid tracking errors being caused by slightly distorting the tracks. Special sections within the track are reserved for auxiliary information, e.g. for recording programme titles and time codes. The auxiliary information capacity of 273 Kbit/s of the R-DAT standard is about four times larger than that of the CD. Additional identification codes are therefore provided, containing information about the sampling frequency (48/44.1 or 32 kHz), number of channels (two or four), type of quantisation (16 bit linear or 12 bit companded), tape speed, copying protection, pre-emphasis and extra information (Table 2.2).

Modes	I	II	III	IV	V
Channels	2	2	2	2	4
Sampling rate (kHz)	48	44.1	32	32	32
Quantisation					
linear	16	16	16	-	-
non-linear	-	-	-	12	12
Subcode (kbit/s)	273.1	273.1	273.1	136.5	273.1
ID-CODE (kbit/s)	68.3	68.3	68.3	34.1	68.3
Revolutions/s	..........................2000...
Playing time	....................2 h/80 min. ....
Tape speed	....................8.15 mm/s ....

Table 2.2 Operating data of R-DAT system [41]

2.1.3 MOD (magneto-optical disc recorder)

The MOD recorder uses a new technology where data are recorded for reproduction on magneto-optical discs. This technique is primarily intended for use in audio recordings to attain the quality of the CD. With appropriate data reduction, recording of digital video signals can also be acheived, in principle.

A thermomagnetic method is used for recording, i.e. a laser in conjunction with an external magnetic field. For reproduction, the magneto-optical Kerr effect is used (effect on amplitude and phase of a light beam when reflected from a metallic mirror). For re-recording, the previously recorded data are directly overwritten by MFM (magnetic field modulation) [36].

Ninety-nine programmes can be directly addressed via an automatically generated table of contents. Without data reduction, the playing time of a disc is 75 minutes. Using the data reduction method MSC (multiadaptive spectral audio coding), specially developed by Thomson, the playing time increases fourfold.

MSC uses a special processing algorithm configured in a VLSI component. The compressed signals are produced by means of a discrete Fourier transform and a special filtering process which corresponds to the response of the human ear. The multichannel signals (subbands) are analysed and efficiently encoded, so the original PCM bit stream is reduced by a factor of four. The MSC decompression during reproduction produces 16-bit PCM output signals which are then processed for acoustic reproduction.

Recording	Thermomagnetic recording
Format	Eight-to- fourteen modulation (EFM)
Reproduction	Optical Kerr effect/evaluation of polarisation change (±0.4º)
Error correction	CIRC
Laser	Wavelength 830 nm, power 6 mW erase, 1.3 mW read
Tangential velocity	2 m/s (CLV)
Coding	DFT algorithm
Quantisation	16 bit (input and output)
Bit rate	≥ 120 kbit/s per mono signal
Disc	5"
Recording layer	2 x 1.2 mm polycarbonate with impressed, spiral-shaped track active layer: amorphous Tb-Fe-Co-alloy
Capacity	450 Mbyte
Playing time	45 min/side

Table 2.3 Specification of the MOD system [36]

In contrast to the CD, where data are read out in the form of a change in reflection, the change in magnetisation is detected by the magneto-optical method. During recording, the magnetic layer is heated by a modulated laser beam. At these points, the direction of magnetisation of the layer is parallel with an external magnetic field. Differently magnetised pits are produced which cause a rotation of the polarisation plane of the laser beam during operation. This change in polarisation is decoded by optical circuits recovering the binary data thus recovered (Table 2.3).

2.1.4 DCC (digital compact cassette)

The DCC developed by Philips and Matsushita was ready for marketing in 1992. DCC is a kind of DAT recorder which is already in use on the professional and semiprofessional markets. The quality of sound from the DCC is comparable to that of the CD and DAT, but the main advantage of DCC is its partial compatibility with the analogue compact cassette. This is an essential criterion since families in the West on average own more than 50 compact cassettes either in their home or car [35]. This compatibility criterion means that DCC has to use the longitudinal recording process, with a tape speed of 4.75 cm/s. These conditions alone require considerable data reduction.

The DCC method employs PASC (precision adaptive subband coding) and is a subset of ISO Layer I (Section 2.4.4). With PASC, the data rate is reduced to 384 kbit/s per stereo signal. Similar to the MUSICAM method described in Section 2.4.2, the frequency dependence of the audibility threshold of the human ear is used. The masking effect, with weak signal frequencies being masked by stronger ones, is also used for data reduction. For utilisation of the so-called resting threshold and the masking threshold the audible frequencies of the audio signal are subdivided into 32 subbands. For each subband, the two thresholds are determined on the basis of the signal level in each band and relating to the signal level in the adjacent band. Single-band signals above the determined dynamic audible threshold (resting threshold plus masking threshold) are quantised with the resolution required for that particular subband. Floating-point conversion is used for this purpose, the mantissa corresponding to the resolution (2 to 15 bits) and the exponent stating the scale factor. The masking threshold and the scale factor are determined for a PASC frame containing 12 successive sampling values.

With DCC, the PASC data stream of 384 Kbit/s is complemented by data for error correction and synchronisation yielding a recording data rate of 768 Kbit/s. As a protection against dropouts on the magnetic tape, the PASC data are distributed over the eight data tracks of the whole tape frame. The additionally recorded bits are provided for error detection and correction using the cross-interleave Reed-Solomon code. Data are being interleaved and distributed among eight data tracks. The CIRC system allows up to four symbol errors to be detected and corrected within one block. The error correction systems can compensate for dropouts of up to 1.45 mm in diameter, corresponding to a total failure of a data track. Uncorrectable errors are concealed [42].

Each of the eight data tracks of a tape holds a data stream of 96 Kbit/s. A ninth track holds 12 Kbit/s of track numbers, start codes and the time code for the counter. The DCC is designed as an autoreverse system, so that a DCC tape has 18 tracks in total. For playing analogue compact cassettes, the head has two additional 0.6 mm wide analogue tracks.

A DCC machine with 2 × 45 minutes playing time was been introduced late in 1992. Its frequency range was determined by the sampling frequency: 48 kHz sampling frequency for the frequency range to 22 kHz, 44.1 kHz to 20 kHz and 32 kHz to 14.5 kHz.

DCC cassette tapes manufactured by BASF comprise chromium dioxide needles with a density of six billion particles per square millimetre of the tape surface [43].

2.1.5 MD (mini disc)

In Japan, late in 1992, the Mini Disc was launched with the MD system combining both CD and MOD techniques. The MD storage medium is an optical disc with a diameter of 6.3 cm built into a dustproof 2.5 inch plastic cartridge giving 74 minutes playing time and an almost unlimited lifetime.

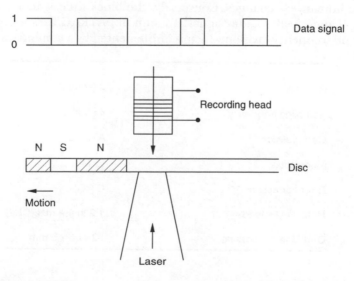

Fig. 2.6 Principle of MD

The optical disc uses ferromagnetic materials whose optical characteristics suddenly change as a function of the magnetic polarity (Fig. 2.6). The information is contained in the shift of the polarisation plane of the laser beam. For storing the data, the terbium ferrite cobalt material is heated to the specific Curie temperature of 489 K [35]. Under these conditions, the polarity of the material can be reversed by applying an external magnetic field. Binary states are stored as magnetic north and south poles. The substrate is heated by a laser, which is also used for sampling.

The discs contain a user TOC (table of contents) in which the playing order, start and stop times of the recorded content are registered. A manufacturing technique makes the MD resistant to impacts which may cause momentarily skipping in the read process. With the aid of cyclically stored address data at 13 ms intervals, the read head can be repositioned to the address last selected to avoid audible dropouts that may occur, while a buffer memory of 1 Mbit (subsequently 2 Mbits) is used, to ensure programme continuation for three seconds [44, 45].

The MD uses the data reduction method ATRAC (adaptive transform acoustic coding) which is similar to the PASC method and also the psychoacoustic effects of masking and audibility threshold. With the aid of spectrum analysis the audible spectral components are filtered out. The audio signal is divided into time blocks and analysed block by block. For each block, the signal is divided into frequency bands whose width, according to psychoacoustic findings, increases with the frequency. At the lower frequencies, subband analysis is more complex than at the higher frequencies. ATRAC also uses a time block length that is not constant. With a large change in the signal, the block time is reduced since the sensitivity of the human ear changes. Conversely, the block time is increased with a slight change in the audio signal. To sum up, ATRAC uses a nonconstant time and frequency division which is implemented by a combination of filter

Channels	2 (stereo)
Sampling frequency	44.1 kHz
Compression	ATRAC
Modulation	EFM
Error correction	CIRC
Rotational speed	1.2 to 1.4 m/s (CLV)
Cartridge dimensions	72 x 68 x 5 mm

Table 2.4 Technical data of MD

and transformation circuits. MD sound assessment tests have revealed that differences from the original sound can be detected: high pitches sound harder, exhibit less reverberation, have a coloured noise signal and are less transparent [35]. The severity of these defects depends on the musical content of the test piece and may not affect the overall listening pleasure (Table 2.4).

2.2 RDS (radio data system)

In 1984 the RDS system was introduced as a further development of VHF FM broadcasting. A subservice of RDS is a traffic information service for motorists called ARI (Autofahrer-Rundfunk-Informationen), which operated in Germany and the neighbouring countries prior to the introduction of RDS. RDS uses a free space in the stereo multiplex signal at 57 kHz for a 1.2 Kbit/s wide data channel. In line with the original intention, this channel carries specific data for improving the sound broadcasting service. The basic idea of RDS to provide additional information for mobile reception is continued in DAB.

Ever since 1976 the various member countries of the EBU (European Broadcasting Union) have proposed a total of five systems for the transmission of radio data in VHF sound broadcasting. These systems were tested in several European countries (for example in Switzerland, UK and Sweden). Common to all systems was the transmission of the information on auxiliary carriers in the stereo multiplex signal, the organisation of the data in blocks and a transmission rate of about 1.2 Kbit/s. Evaluation of the test results clearly showed the superiority of the Swedish system so in 1992 this system was chosen by the EBU as the standard European system [46].

Today, RDS has been adopted in almost all European countries and is being introduced in many non-European countries. In the US, RBDS (Radio Broadcast Data Service) is considered to be the most important innovation for the radio receiver in the past decade. In March 1993, NRSC (National Radio Systems Committee) passed the RBDS standard which is based on the European RDS. An adaptation to MF sound broadcasting is being developed.

2.2.1 Information transmitted by RDS

Within RDS, tuning aids, switching signals, radiotext and various other information services can be transmitted (Table 2.5):

Tuning aids

PI (Programme identification)
PI identifies transmitters in a programme service with identification bits for

the country, the area (corresponding to the traffic broadcast area of the ARI system), and the programme service (e.g. broadcasting corporation and programme number). Receivers are able to scan and retune automatically to a transmitter broadcasting the programme service.

PS (programme service name)
The PS is indicated by eight ASCII characters on the radio receiver's display.

PTY (programme type)
Identification bits for scanning receivers for the selection of various types of programmes such as news, sport or classical music.

TP (traffic programme identification)
Identifies transmitters broadcasting traffic information (corresponding to the 57 kHz carrier in the traffic information system).

AF (alternative frequencies)
Lists the frequencies of adjacent transmitters broadcasting the same service. In mobile reception under adverse conditions it is possible to switch to an alternative transmitter.

Switching signals

TA (traffic announcement identification)
TA is equivalent to the announcement feature in a traffic information system. The switching signal can be used for automatically switching to a traffic announcement from, for instance, the cassette player, a muted receiver, or a programme being received without traffic information (but in the latter a second receiver module is required).

DI (decoder identification)
Switching signal for operating mode selection, e.g. mono, stereo (previously identified by a 19 kHz pilot), dummy stereo head, high-com.

MS (music/speech identification)
Identification of speech or music broadcasts for individual volume and tone control.

PIN (programme item number)
Transmission of programme time and date as published in the magazines to enable automatic switch-on of the receiver or recorder, with any re-scheduling of programmes automatically allowed for.

RT (radiotext)
Information accompanying the programme (e.g. title of music, name of interpreter, programme change) is transmitted in the form of text with 32 (maximum 64) ASCII characters for display on home receivers, but, for safety reasons, in-car receivers use a speech synthesiser (voice coder).

Further applications

ON (other networks information)
Various identification signals are transmitted for up to eight other programme services: PI, PIN, TP, PTY, TA, AF.

TDC (transparent data channel)
This may be used to open up to 32 data channels with channel number and start and end criteria of the data files. The transmitted alphanumeric and graphic characters are suitable, for instance, for screen display similar to teletext in television.

IH (in-house application)
This refers to data decoded only within the broadcasting organisation. It is possible to transmit the identification of programme origin, transmitter-related switching signals, data for monitoring audio quality or radiopaging calls.

CT (clock time and date)
To avoid ambiguities in the reception of radio data from different time zones or due to differences in summer and winter times, the date is broadcast according to MJD (modified Julian day) (days counted from 1st March 1900) and the time according to UTC (co-ordinated universal time) (hour, minute, local offset), while conversion to local time and date is made through the RDS decoder.

	Group	Bits per message	Capacity % Σ 100%
PI	all	16	25.0
PS	0	72	9.4
PTY	all	5	7.8
TP	all	1	1.6
AF	0	8	8.3
TA	0	1	0.52
DI	0	4	0.52
MS	0	1	0.52
PIN	1	37	4.8
RT	2	296	9.6
ON	3	37	4.8
TDC	5		14.5
IH	6		4.8
CT	4	37	
GA	all	5	7.8

Table 2.5 Information carried by RDS (GA = group address)

2.2.2 Coding and modulation

2.2.2.1 Coding of radio data

Radio data are transmitted in the form of a continuous, binary data stream with 1.1875 Kbit/s. The data organisation provides for groups of 104 bits, each group comprising four blocks of 26 bits each (Fig. 2.7). The blocks contain the 16-bit information word and a 10-bit checkword. Through the integration of offset words and Boolean operation (modulo-2 division) with a generator polynomial, the checkword can be used for block and group synchronisation and for detection and correction of transmission errors [47].

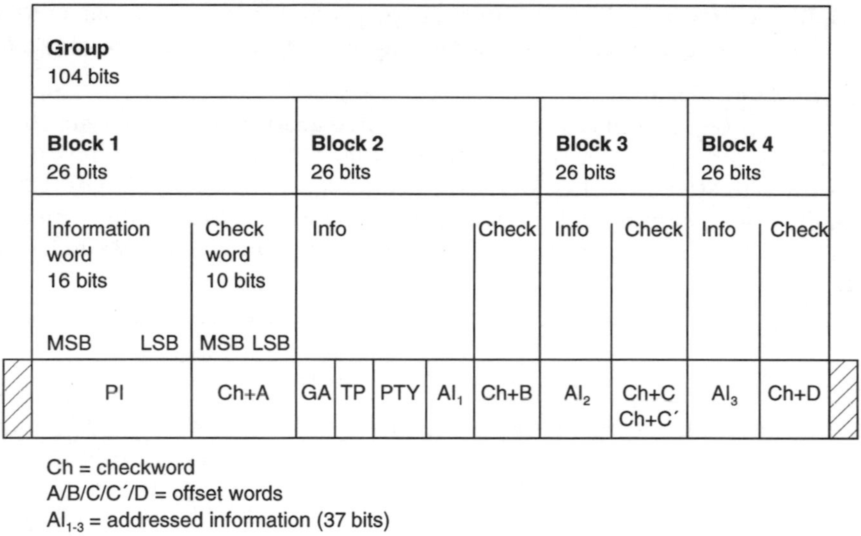

Ch = checkword
A/B/C/C′/D = offset words
AI_{1-3} = addressed information (37 bits)

Fig. 2.7 Structure of RDS data

The type of group is defined by the group address GA with 5 bits, but with only 16 possibilities being used. One bit is used to distinguish between the two group versions A and B (in version B the PI code is repeated in block 3 for faster search tuning). The first block of each group always contains the PI code for fast programme identification and the second block the group address, TP (traffic programme identification) and PTY (programme type). The remaining capacity in blocks 2 to 4 will be occupied according to the type of group. The sequence of the individual group types can be selected at the transmitter end. It depends on the importance of the message to be transmitted, with fixed, minimum repetition rates to be considered for certain information.

In selecting the modulation carrier and type of modulation for the RDS signal, the existing occupancy in the stereo multiplex baseband had to be considered particularly. Fig. 2.8 shows the audio midband signal (to 15 kHz) and the stereo pilot tone (at 19 kHz), the sideband signal (23 to 53 kHz) and the ARI signal with station identification, area identification and announcement identification (at 57 kHz). The ARI signal is amplitude modulated with a modulation depth of 60% for the area identification and 30% for the announcement identification. Due to the low modulation frequencies, the frequency spectrum has a very narrow bandwidth (maximum 57 kHz ±125 Hz with announcement identification switched on). The broadband RDS signal is superimposed on this ARI signal. The RDS should on the one hand enable a high data rate and on the other hand not affect the VHF programme signal and the ARI signal for compatibility reasons. This means that the frequency spectrum of the RDS signal should have no significant components at 57 kHz and around 53 kHz. Optimally, the maximum power should be in the middle of the frequency gap. The same applies to the frequency range above 57 kHz, where the frequency components of a VHF data transmission link, the so-called 60 kHz audiodat system, lies at 60 kHz [48].

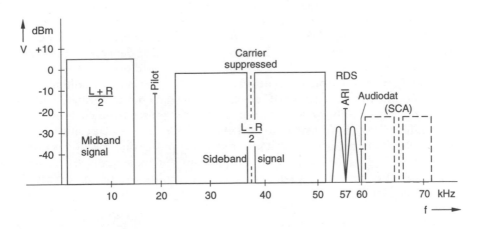

Fig. 2.8 Stereo multiplex baseband with RDS signal SCA (subsidiary channel authorisation)

These above conditions can be met by the special processing of the RDS data stream to limit the frequency spectrum to ±2.4 kHz and subsequent double sideband modulation (DSB-AM) with carrier suppression. Moreover, the RDS carrier is phase-shifted by 90° relative to the ARI carrier (provided that the ARI carrier is being transmitted) so this so-called quadrature modulation prevents the ARI signal from being unduly affected.

The RDS spectrum within the multiplex signal is shown schematically in Fig. 2.8. The RDS data rate of 1.1875 Kbit/s is locked in integral numbers to the 57 kHz carrier and also to the 19 kHz pilot tone. This locking to the stereo pilot prevents adverse effects on the stereo signal. The multiplex signal with RDS added is frequency modulated by the VHF FM transmitter. The peak deviation of the transmitter with 100% modulation should not exceed ±75 kHz for the composite multiplex signal. A maximum of 10% thereof, that is ±7.5 kHz, is available for the additional RDS and ARI signals. Field tests have shown that a deviation of ±3.5 kHz for the unmodulated ARI carrier and of ±1.2 kHz for the RDS signal is the best compromise. In addition to quadrature modulation, these deviation levels prevent impairment of the ARI signal while ensuring sufficient data transmission reliability, in particular around the boundaries of the coverage area [48].

2.2.2.2 Biphase data modulation

As illustrated in Fig. 2.7, the RDS information is combined into groups of 104 bits and protected by an error detecting code. The various groups are applied to the modulator as a binary serial data stream and the data stream is modulated in several steps (Fig. 2.9). The binary data are differentially coded by a logic 1 at the input, causing a level change at the output and a logic 0 at the input to maintain the output level. This enables any signal inversions in further processing to have no effect on the data decoding the receiver, since the RDS decoder performs differential decoding so regenerating the original data. The differentially coded data are not suitable for direct modulation at 57 kHz since high power would occur at the carrier. With a larger number of consecutive logic zeros the receiver would also no longer be able to recover the RDC clock from the data signal. Biphase coding avoids this problem by generating a pair of Dirac pulses from each bit. The logic information is transmitted with the pulse pair, and a positive Dirac pulse followed by a negative, or a negative Dirac pulse followed by a positive pulse. This implies that even with longer constant bit sequences the RDS clock will also be transmitted. Moreover, no significant power occurs at around 57 kHz [49].

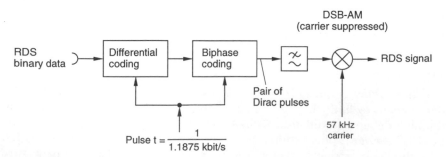

Fig. 2.9 Principle of RDS data stream processing

In the EBU specifications, the double Dirac pulses have a spacing of half a bit duration T_D. Since theoretically these pulses have an infinitely wide frequency spectrum, the spectrum bandwidth is limited to 57 kHz ±2.4 kHz by filtering. The amplitude response of this filter defined in the EBU specifications corresponds to a 100% cosine roll-off in the range from 0 to 90°, with the cutoff frequency being 2.375 kHz.

As result of pulse filtering, time functions of the biphase symbols are obtained in accordance with the bit sequence. Note that, due to bandlimiting, overshoots of theoretically infinite duration occur at both ends which are superimposed on the waveforms of the adjacent biphase symbols. Fig. 2.10 shows the generation of a biphase signal for the bit sequence 10011010100. This signal is subsequently modulated at 57 kHz and added as an RDS signal to the MPX signal (Fig. 2.11).

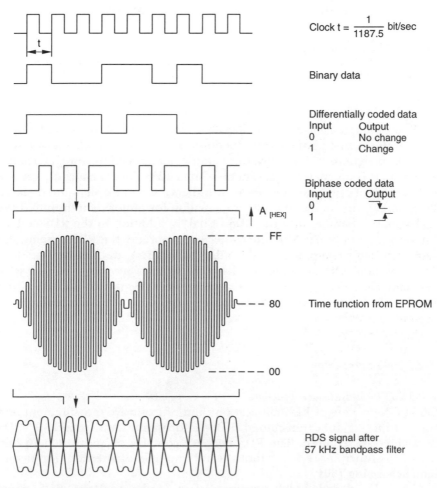

Clock $t = \dfrac{1}{1187.5}$ bit/sec

Binary data

Differentially coded data
Input	Output
0	No change
1	Change

Biphase coded data
Input	Output
0	
1	

A [HEX]

FF

80   Time function from EPROM

00

RDS signal after
57 kHz bandpass filter

Fig. 2.10 RDS pulse diagram

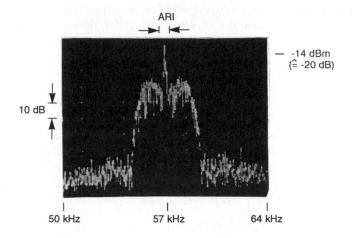

Fig. 2.1 RDS-ARI sum signal
RDS = statistical 0/1 distribution; ARI = area identification F with announcement
identification ON

Radio data transmission was officially introduced in Germany on 1st April 1988, with the first phase of introduction featuring station-related RDS data such as programme identification, station identification and alternative transmitted frequencies. Programme-related RDS data broadcasts, such as radiotexts or programme item numbers, were implemented later. Programmes broadcasting traffic information for motorists transmitted the ARI signal in parallel with the RDS signal. In addition to the RDS coders, fitted with an optional ARI coder where appropriate, the RDS systems also require suitable equipment for decoding, monitoring and transferring data. The RDS (and ARI) decoders are used for monitoring purposes, as system and monitoring decoders for transferring the RDS data to an RDS coder (data link) and as RDS-ARI measuring equipment in measuring and monitoring systems.

2.2.3 RDS monitoring

2.2.3.1 RDS measurement functions
Due to the variety of RDS applications and of extensive measurement and monitoring of the transmitted RDS information and analogue RDS parameters is required. The RDS decoder contains all measurement and evaluation functions required for RDS and a number of interfaces for further data processing [50].

The RDS decoder allows measurement of the analogue RDS signal parameters in the stereo multiplex baseband, i.e. the level of the RDS signal and the phase difference between the RDS signal and the 19 kHz stereo pilot and between the RDS and ARI signal, which is 57 kHz. The measured levels have a resolution of 1 mV or 0.1 dB; phase resolution is 1∞ to within 5∞. Using its decoding function, the RDS decoder retrieves the serial data bits from the differentially biphase-coded double-sideband RDS signal and synchronises to this data stream, which is arranged in blocks and groups. The decoder features a number of error correction methods and with the aid of the checkbits included in the data stream, errors can be detected and corrected, using the error correction facilities specified in the RDS standard. The error detection and error correction modes (1 bit, 2 consecutive bits and 2 bits in a 5-bit interval) can be selected.

PI, PS and TP are the significant data of a programme. These data are checked with the source monitoring function. The decoder has reference memories storing the nominal data for these three characteristic RDS features. If there is no agreement between the activated reference memories and the RDS data received, a fault message is issued.

The RDS data PI, PS, TA, TP, MS, DI and PTY are also checked. Any change to these RDS data is monitored separately to determine whether any of the RDS data change more frequently within a defined period of time rather than the limit specified for such a period. If the limit is exceeded, a fault will be signalled.

The reception quality of the RDS signal can be determined from the block error rate. The block error rate is the proportion of errored blocks within the last 100 blocks received and calculated as a floating average value as a percentage. If one of the error correction modes is selected, the number of corrected blocks is indicated [52].

2.2.3.2 RDS monitoring
For continuous monitoring, an RDS data decoder is permanently connected to an RDS channel, allowing particular monitoring and logging of the dynamic data.

The advantage of cyclic testpoint selection is that a single RDS data decoder can monitor the RDS data of several programme services. The decoder is connected to the various RDS channels, allowing the static RDS information and analogue parameters of the RDS signal to be appropriately monitored. This type of monitoring is based on the assumption that the analogue characteristics of the RDS signal will not change abruptly and the static RDS information, such as switching over to night or regional programmes will not be modified frequently. If these provisos are met, the breaks during which the RDS decoder monitors other testpoints are acceptable. Link and source errors are detected by the RDS monitoring decoder after the monitoring cycle has covered all testpoints (test intervals). Each error is stored with the MPX signal source. Multiplex selectors and tuners are used for connecting the monitoring decoder to the testpoints.

Multiplex testpoints and VHF FM testpoints can be monitored in the same cycle. The number of testpoints and frequencies to be scanned can be selected [51].

Since the individual programmes do not run continuously and therefore do not require round-the-clock monitoring, the decoder monitors RDS channels only during an assigned time window. Source monitoring can be activated independently with the aid of the multiplex selector or tuner for each RDS channel to be checked.

2.2.4 Special RDS applications

2.2.4.1 TMC (traffic message channel)

In RDS the TMC is transmitted in the RDS group 8A with 37 bits being available per group. With about six to eight TMC groups per second (depending on the RDS channel occupancy), about 200 to 300 bit/s are available for TMC. The ALERT-C protocol is tailored to the needs of RDS transmission. ALERT is the acronym for 'advice and problem location for European road traffic'. The ALERT-C protocol is a fixed protocol for the transfer of messages, from the detection of the problem through to the broadcasting studio. ALERT-C is to become a CENELEC (Comité Européen de Normalisation Electrotechnique) standard [54].

Compressed information is transmitted in a neutral data format, i.e. addresses, which are expanded in the national language in the receiver itself. Place, duration and warnings are transmitted, with any additional information, and can be output on paper, a display (memo function) or a vocoder.

The basic conditions for customisation of the traffic information service is possible. After data entry of the journey starting point and destination, the receivers — with sufficient storage capacity — can search the database for relevant information on the planned route and inform the driver in his own language at the push of a button. A thin chip card the size of an identity card not only carries the decoder but also a voice synthesiser with a standard vocabulary, including the geographical names of the respective country. This smart card also determines the language and nationality of the car radio [55].

The TMC data can be routed from the broadcasting studio to the FM transmitter stations via 400 bit/s modems within the frequency range of the modulation line with the following specifications:

- carrier 14.85 kHz
- modulation mode 4(8) — (D)PSK
- level −25 to −30 dBm
- bandwidth 14.75 to 15.00 kHz
- bit error rate max. 10^{-2}.

A possible alternative is a free data channel for auxiliary information within the DS1 signal in DSR (Section 4.4).

TMC transmission is also being investigated in AM systems (AM-AI). Tests are being carried out on MF and LF transmitters with amplitude modulation for transmitting AM auxiliary information with a gross rate of 200 bit/s. Transmission is possible by means of phase modulation of the carrier, with a data structure similar to that of RDS.

In the long run these efforts are expected to lead to a European-wide traffic information system (Fig. 2.12). There are local information centres, one national centre and later there will be a European centre for information processing. From these centres the information is passed via the broadcasting studios to the transmitters (VHF, MF, LF). Highway authorities, police and automobile associations (ADAC in Germany) are integrated into the system.

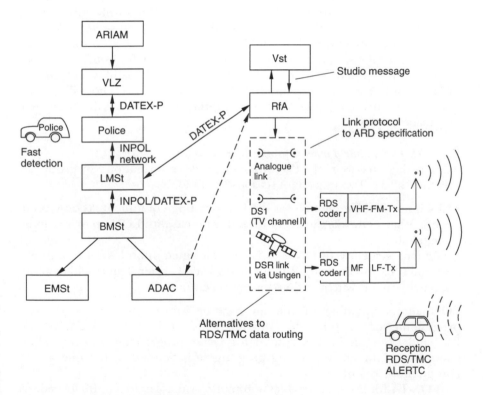

Fig. 2.12 RDS/TMC information flow

ARIAM = autom. traffic sensor/counter
VLZ = traffic control centre

RfA = broadcaster
Vst = Traffic studio

LMSt = local
BMSt = national information centre
EMSt = European information centre
ADAC = German Automobile Association

ALERT C = protocol
format (on message lines
and with terrestrial
broadcasting)

For data acquisition, an automatic traffic data collection system (ARIAM) has been partially installed and is to be expanded to full coverage by 1995.

BEVEI, which stands for 'better traffic information', is a project subsidised by the German Ministry for Research and Technology and implemented under the control of Bosch/Blaupunkt for the German broadcasting services WDR and Südwestfunk. A field test with real TMC data was initiated early in 1993. Regular TMC operation is expected to start in 1995.

The decision to introduce TMC Europe-wide will not only enhance the car radio but also raise interest in the nonmobile sector. Possible applications, for example, could be stationary professional TMC reception/display stations, equipped with large displays for hotels, car parks, airports and railway stations and with dialogue capability (electronic information). For motorists using radios without TMC, who will be the majority, on introduction of the system, another application is the motorway automatic display.

2.2.4.2 EON (enhanced other networks information)
EON is the successor of ON (other networks information) in group 3A (Section 2.2.1). Two important reasons led to EON:

• ON is not flexible enough for the broadcasting equipment, as only up to eight other networks can be addressed, whereas with EON up to 20 can be included.
• The data structure of ON is unfavourable since upon loss of one group several subsequent groups cannot be evaluated either. This would result in a much longer switching time for the receiver.

EON allows updating of the information stored in the receiver via programme services other than the one being received. AF, PS, TA, TP and PTY and PIN can be transmitted for other programme services. Their allocation to the corresponding programme is defined by the relevant PI (programme identification).

EON-TA for instance, enables automatic switchover to a traffic broadcast transmitter for the duration of the traffic announcement. EON-PTY allows switchover to a certain programme type (PTY).

Group type 14, in which EON is transmitted, exists in the two versions A and B. Version A is the normal process used for the background transmission of EON information. The maximum cycle time for the transmission of all data relating to all cross-referenced programme services is less than two minutes, while Version B of group type 14 is used to indicate a change in the state of the TA flag of a cross-referenced programme service. The broadcasting corporations should choose the suitable method for each cross-referenced programme service.

2.2.4.3 EWS *(emergency warning system)*

A number of countries are interested in providing comprehensive emergency information on a national basis, e.g. warnings of adverse weather, chemical accidents or military threat (Fig. 2.13). Codes would be used which for reasons of security cannot be fully detailed. The information would only be broadcast in cases of extreme emergency so as not to overload the RDS channel by these codes. This new service is independent of the already defined alarm code system (PTY=31).

The following identification is required to operate the emergency warning system. Group type 1A is used to identify the programme that carries these warning messages to activate special receivers for processing these messages. The time interval for broadcasting the warning messages depends on the national implementation. Normally, the time interval for one group of type 1A should not be shorter than two seconds.

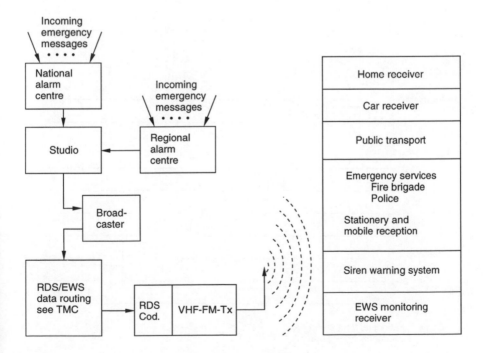

Fig. 2.13 EWS (emergency warning system)

2.2.4.4 RP (radio paging)

Radio paging is a system for paging calls to a specific receiver/person by transmitting information via the pager's 6-digit code address.

The transmission sequence of the paging system is time controlled and for this reason a timing pattern is produced with group 1A. To produce the timing pattern, group 1A need not be entered, but is automatically inserted into the group sequence every second. Every full minute group 1A is replaced by group 4A (CT) [56].

The time control in the coder supports the battery saving mode of the paging receivers (Fig. 2.14). Each minute is divided into ten equal intervals (intervals 0 to 9) of six seconds. In each interval, only those paging calls are transmitted whose last digit of the paging address is identical to the interval number. The paging receivers synchronise to the timing patterns in the coder and only become active in the intervals in which a message can be transmitted to them. The pager with the number 123456, for instance, is activated in the sixth interval only (36th second after a full minute). If no paging information is received within the next ten groups, the pager is inactive again, but if a message is transmitted it may also cover the next interval and the next interval after that.

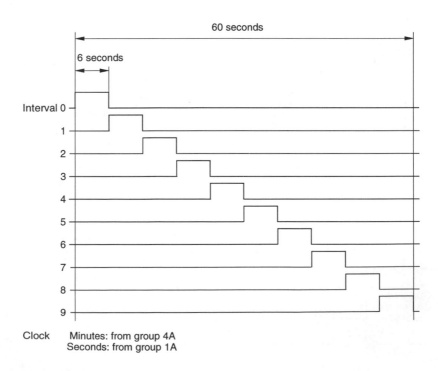

Fig. 2.14 Radio paging transmission sequence

The message formats for paging calls are transmitted in group type 7A:

- paging without additional message
- paging with additional numeric message, 10-digit
- paging with additional numeric message, 18-digit
- paging with additional alphanumeric message, maximum 76 characters.

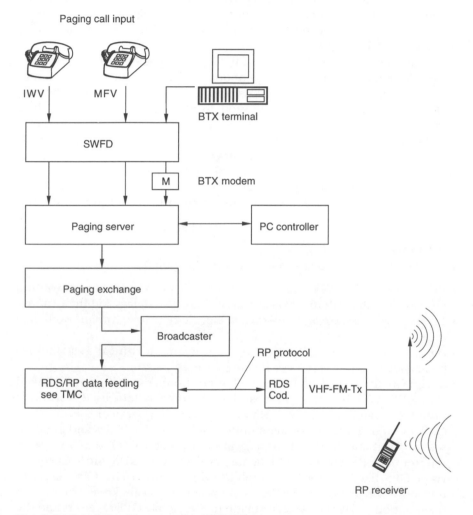

Fig. 2.15 Principle of an operative RDS/RP system

IWV = pulse dialling method
MFV = multifrequency method
SWFD = subscriber trunk dialling of German Telekom
BTX = videotex.

In the RDS coder, ten interval memories (queues) each of 70 group 7As are provided. The paging calls are sorted according to their last digit and placed into these queues. They are transmitted at the moment the receivers are activated and decode the data stream.

The RP method can be used to designate groups of pagers, the so-called transmitter network groups. The first two digits of the paging address are the group code. Each group has a certain number of pagers. The group designation makes it possible to distribute the paging calls to several networks (transmitters), e.g. several networks during the daytime and a single network during the night.

Within the radio paging system of the RDS the paging call is transmitted via public networks, similar to telephone and fax, to the paging centre (Fig. 2.15) where the paging exchange and paging server processes the calls and transfers them to the transmission line of the VHF FM transmitter station. The protocol on these lines is based on the RDS standard specifications [52], and TNPP protocol (telocator network paging protocol) being used internationally. Both protocols are bidirectional, i.e. with a command and reply. The call is transferred via the RDS coder in the transmitter station. Paging receivers are available from NOKIA (main supplier, in Scandinavia, France and other countries) and Mitsubishi and AXCESS (USA).

2.2.4.5 Other future applications
There are many other plans to use the versality of RDS:

- *information for public transport*: visual displays showing travelling information about destination and arrival times or delays and their causes;
- *banking and stock exchange information*: e.g. share prices, foreign exchange rates, interest rates;
- *control and monitoring systems:* for automatic technical installations controlling pumps, motors or heating systems;
- *navigation*: the differential GPS (global positioning system) can be linked with RDS. GPS is a navigation and positioning system used worldwide utilising 21 satellites and SPS (standard positioning service) specification. Civil and military users locate, monitor and control vehicles and goods at sea, on land and in the air to a location accuracy of 20 to 100 m. With differential GPS, correction data are received from RDS and transferred via a GPS fixed station to a mobile GPS receiver. The GPS data are transmitted in group 3A of RDS. As the position data are known to the GPS fixed station, corrections can be made in the mobile GPS receivers and the accuracy of positioning considerably increased (to within 2 m) [53, 57].

The large variety of possible RDS applications has created the need for a second or third RDS channel. Since the SCA channel has not widely been introduced in Europe, this channel could be for further RDS channels. Carrier frequencies of 63 kHz and 67 kHz in the multiplex signal have been considered.

2.3 DSR (digital satellite radio)

Although it was a particularly powerful system, DSR was ill-fated. It was developed in the mid-1980s by the IRT in co-operation with the German DBP-Telekom and it was initially planned to install and introduce the system on the direct broadcasting satellite TV-SAT1. The loss of TV-SAT1 and the failure to provide an alternative transponder — for obvious reasons of audio being given less importance than video — led to several year's delay in the introduction of DSR. In the beginning, DSR was considered purely as a German system, which made it difficult for it to gain acceptance throughout Europe. Moreover, potential manufacturers of home receivers were rather reluctant to manufacture units to sell below £450.

DSR finally made the breakthrough in 1991 at the International Consumer Electronics Show in Berlin. Since then DSR has been transmitted by the two satellites KOPERNIKUS and TV-SAT2 with practically all the programmes fed into the cable network of German Telekom. The products are now priced acceptably to the consumer.

There is one great disadvantage of DSR that can probably not be solved economically — mobile reception. There are research projects on mobile DSR reception using controlled phase array antennas. These antennas are computer controlled for alignment to receive other signals from the individual antenna array elements. However, due to the high manufacturing costs, these antennas are not seen to be a possible viable consumer product.

A further disadvantage of the system seen by DSR subscribers is that it only has the capacity for 16 stereo programmes, although this disadvantage can be overcome. The introduction of data reduction in baseband coding, e.g. using MUSICAM, can increase the DSR capacity by a factor of four without loss of quality both in an uplink and downlink compatible solution (Section 4.1) making DSR more attractive for reception via cable or satellite.

There is no problem regarding the number of programmes offered, as now over 100 programmes are provided by various satellites in different modes of modulation and coding.

Finally, DSR features a characteristic which can be utilised in many ways as described in Chapter 4. DSR provides a transparent gross data capacity of 20.48 Mbit/s which may be used for professional purposes, e.g. not only for programme distribution, but also consumer applications covering both home receivers and portables.

In the following section the underlying principles and techniques of DSR are explained.

2.3.1 Overview

Digital stereo sound broadcasting via satellite under real operating conditions had its world première in 1985 at the International Consumer

Electronics Show in Berlin. On this occasion, the digitally coded stereo signals were transmitted in a quality that set new standards for sound broadcasting. Sixteen DS1 sources were combined in the earth station of DFVLR (now the German Aerospace Research and Development Establishment) at Oberpfaffenhofen near Munich and transmitted to Intelsat V positioned over the Indian Ocean. In Berlin a satellite receiving system of DFVLR picked up the digital sound broadcast signals and fed them into the cable network of the exhibition centre where several manufacturers were exhibiting digital sound broadcast receivers. It was the first time that trials on digital transmissions were made successfully with sound quality comparable to the compact disc [59].

Compared with the conventional system, digital sound broadcasting via satellite offers three essential improvements to the listener [58]:

- reception of stereo programmes in compact disc quality;
- stationary reception is ensured anywhere in Germany and adjacent areas of neighbouring countries;
- auxiliary information for programme type identification, allowing a programme to be preselected and tuned in automatically by the listener.

In Germany, digital sound broadcasting was established via the earth station of the German DBP-Telekom at Usingen near Frankfurt, the direct broadcasting satellite TV-SAT2 and the communications satellite KOPERNIKUS. Arianespace at Evry near Paris is responsible for the European launch vehicle Ariane. Digital sound programmes are offered by ARD and private programme providers. From the studio output, the baseband programme signal is transferred to the DS1 coder which converts it into a serial data stream in accordance with the DS1 specification. Corresponding to the 2 Mbit/s data hierarchy of the DBP, two such DS1 signals are transmitted in a DS2 channel via the digital modulation line network of the DBP to the earth station at Usingen (Fig. 2.16). A data multiplexer interleaves 16 DS1 signals and, together with frame sync. words, produces two 10.24 Mbit/s data streams, which are applied via a scrambler to the modulator furnishing the QPSK modulated IF carrier. In the satellite transmitting equipment the QPSK signal is converted into the transmit frequency in the band 17.7 to 18.1 GHz. The TV-SAT2 satellite transmits at frequencies between 11.7 and 12.1 GHz, providing full coverage of Germany with three TV signals coded in accordance with the D2-MAC packet system and 16 digitally coded stereo signals.

The satellite broadcast receiving equipment for direct reception of digital sound (and TV) signals comprises outdoor and indoor units. The outdoor unit comprises a parabolic antenna and a 1 GHz downconvertor (950 to 1350 MHz). The 1 GHz signal is used for individual reception and small community antenna systems, otherwise the frequency is converted to 118 MHz in the indoor unit by a convertor connected ahead of the receiver. In addition to the direct reception of TV-SAT signals via individual or

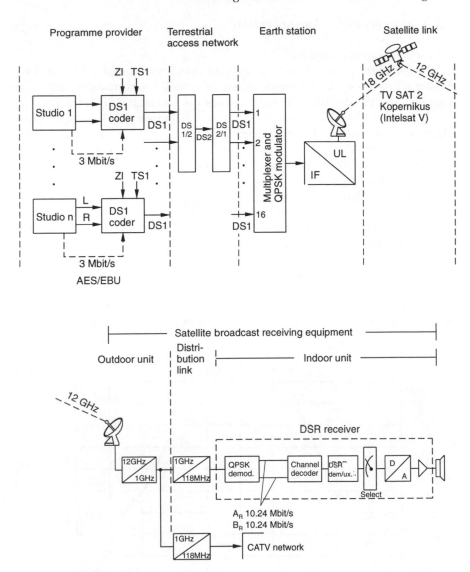

Fig. 2.16 Digital satellite radio transmission link

community antenna systems, the reception at a cable headend and signal distribution via the broadband communication network is of great importance [63]. DBP-Telekom has assigned two special TV channels (S2 + S3, 118 ±7 MHz) for direct feeding of the 14 MHz digital QPSK signal to the receiver.

2.3.2 Principles and techniques of DSR

2.3.2.1 DS1 interface standard

International broadcasting authorities have jointly specified the standard for the digital stereo signal to ensure high quality audio transmission within the framework of the digital microwave links of the German DBP-Telekom. The data rate of 1.024 Mbit/s results from this microwave link compatibility, which involves phase locked coupling of all signal sources. A digital microwave channel which uses a data rate of 2.048 Mbit/s is able to transmit two DS1 signals. The frequency of the DS1 system clock is an integral multiple of the audio signal sampling frequency of 32 kHz. By using the pseudoternary HDB3 code for the output signal, the system clock can easily be retrieved from the latter for decoding [65].

An essential point of this standard is the stipulation for the conversion of the linearly quantised sampling values into a floating point format with a 14-bit mantissa and a 3-bit exponent derived from 64 sampling values.

The 64 samples are combined to form a block and the sample with the greatest magnitude is determined for each block (Fig. 2.17). This value is looked up in the scale factor table shown in Fig. 2.18 (16/14-bit floating point technique). Scale factors from 0 to 7 can be found this way.

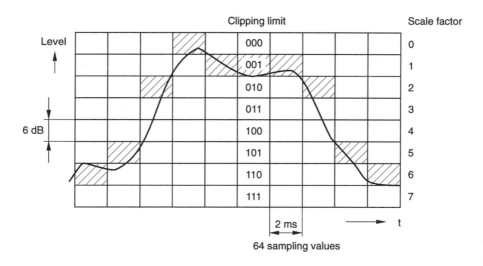

Fig. 2.17 Scale factor derived from DSR audio signal

In the coder, the n most significant bits are clipped off from the 64 samples of a block, since they share the same value and carry no information (Fig. 2.18 on the right).

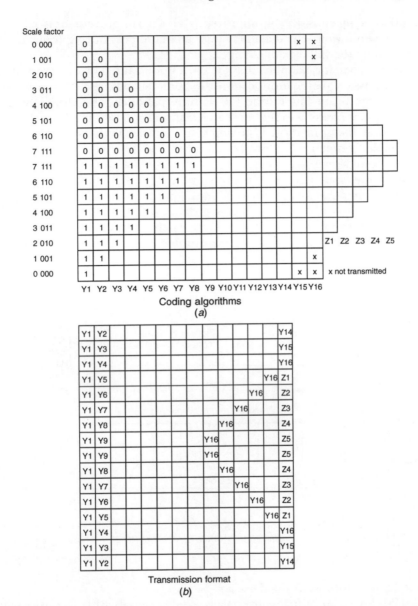

Fig. 2.18 (a) Coding before 16/14-bit floating point arithmetic (b) Transmission format of DS1 signal

Due to this data reduction, which went unnoticed by the listener during trials, space is made for auxiliary information accompanying the programme — similar to RDS in VHF sound broadcasting — and the S/N ratio in DBS reception being considerably improved [64]. This improvement is achieved by 'quasi-windowing' the dynamic range so that errors during transmission do not affect the full dynamic range.

The result is, for example, that the click noise caused by bit errors along the transmission path cannot drown the audio signals, since the error bits are least significant bits during low volume passages. Therefore, the DS1 system is more pleasing to the ear than a transmission system which does not use the scale factor method. Listening tests have shown that no difference is perceptible between a uniformly quantised 16-bit signal and a 16/14-bit floating point signal.

Since the scale factor remains constant for 64 samples, it can be transmitted in serial form, concealed in the parity bit accompanying each of the 14-bit codewords, without affecting data protection. Together with the frame identification required for receiver synchronisation and the auxiliary information, the structure of the transmitted pulse frame is as shown in Fig. 2.19

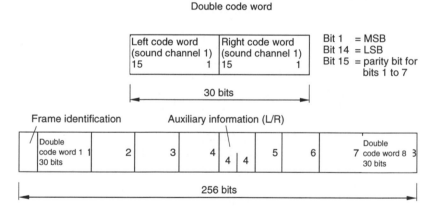

Fig. 2.19 Digital 1024-kbit/s stereo signal (DS1). Top: double codeword structure 32 kHz. Bottom: pulse frame structure with frame clock of 4 kHz

2.3.2.2 DS1 line transmitter and receiver
The DS1 line transmitter converts two audio signals with the associated auxiliary information into a 1.024-Mbit/s serial data stream in accordance with the DS1 specification. There is a need for a sampling rate conversion from 48 kHz (AES/EBU) to 32 kHz (DS1) with the same accuracy.

In the clock generator of the DS1 coder, all further clock signals are derived from clock signal TS1 (1024 kHz) (Fig. 2.20). If the external synchronisation fails, the 1024 Hz system clock is produced internally with crystal accuracy.

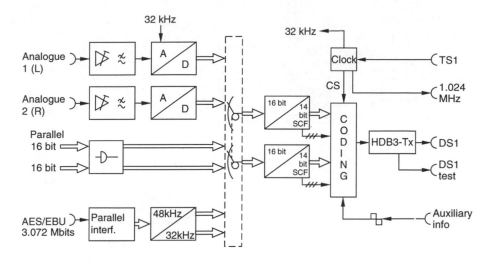

Fig. 2.20 Block diagram of DS1 line transmitter

The frame generation is the central function of the DS1 line transmitter. Here, sync. words, samples and auxiliary information are combined in accordance with the coding method stipulated by the DS1 specification and output as an HDB3-coded data stream.

The digital input allows processing of external digital samples applied in parallel 16-bit format after linear quantisation. The following modules are required for each audio channel and are therefore duplicated in the DS1 coder: in the analogue section the audio signal passes through a variable amplifier and a 15 kHz lowpass filter provided for the suppression of aliasing effects. This filter is a 13th-order Cauer lowpass filter designed by CAE (computer aided engineering) programs. The subsequent A/D convertor produces linearly quantised 16-bit samples at a 32 kHz clock rate.

The compressor converts these samples into the floating-point format with a 14-bit mantissa and 3-bit exponent. Sixty-four samples are combined in a block, the highest of these 64 values determining the exponent.

In addition to its application in digital satellite radio, the DS1 line transmitter is also useful for the distribution network for sound broadcast studios. In this example, the exchange network between the studios is digitised by DS1 line transmitters. Also, they are useful at distribution points of programme providers using digital audio coupling devices fitted with DS1 inputs [59].

The DS1 line receiver outputs the two audio channels in analogue form, the auxiliary information and the status or error messages (Fig. 2.21).

In the clock generator, the data bits and the data clock are regenerated from the DS1 signal. In the decoder, two signal processors evaluate the data

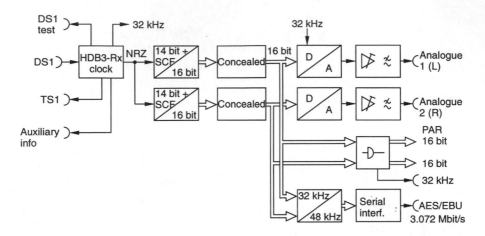

Fig. 2.21 Block diagram of DS1 line receiver

bits from the clock generator. This includes synchronisation to the data stream and continuous monitoring of synchronism as well as output of the auxiliary information via the AI interface and finally the processing of the sound signal samples. The analogue sections perform, separately for each channel, the D/A conversion of the 16-bit samples processed by the decoder, ensuring excellent crosstalk attenuation of more than 90 dB [62].

The DS1 line receiver is also used as a test and monitoring receiver in DS1 lines (Fig. 2.22). The sound quality at the analogue output is monitored by the conventional method. Counter outputs are provided for the determination of parity errors. In addition, the receiver is fitted with a port for the output of messages, such as synchronisation error, HDB3 code error and parity error.

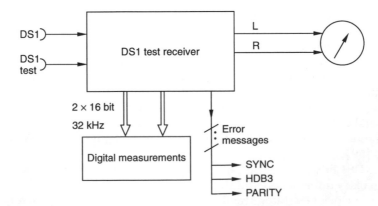

Fig. 2.22 DS1 test receiver

Note: In 1989 the DS1 technique was adopted for the digitisation of the programme exchange network by ARD (Federal German broadcasting corporations) (Fig. 2.23). The DS1 signal is transmitted from the broadcaster via a microwave channel to the distribution point in Frankfurt. From there the signal is transferred to its destination where the DS1 line receiver retrieves the analogue sound signals and the auxiliary information. The transmission system is fully synchronous, since the clock of the DS1 line transmitter synchronises with the line clock of the German DBP [71].

Through the use of MUSICAM and the methods proposed by DSRplus (Section 4.1), the capacity of the line network can be increased by a factor of four.

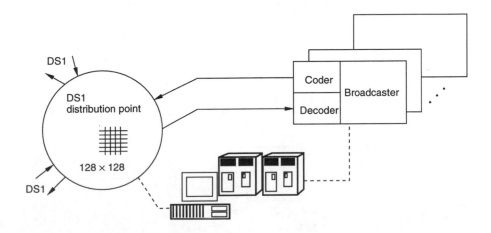

Fig. 2.23 DS1 audio line network

2.3.2.3 DSR multiplexer and modulator

The essential part of the DSR transmission link is the multiplexer/QPSK modulator, which produces a 118 MHz carrier signal modulated with 16 stereo channels. This carrier signal complies with the specifications for digital satellite radio (Fig. 2.25). The digitised audio signals are combined in the two main frames A and B, which contain a number of subframes, to form two 10.24 Mbit/s data streams. The two digital signals are accessible via loopthrough connectors A and B. After differential encoding and pulse shaping, the data streams are routed to the QPSK modulator, which modulates them onto the 118 MHz carrier. The principle of QPSK modulation is illustrated in Fig. 2.24. The QPSK modulator adjusts the phase such as prescribed by the two simultaneously present data bits from the main frames A and B. In practice, phase adjustment is implemented by two two-phase modulators whose output signals are taken via an adder network. The

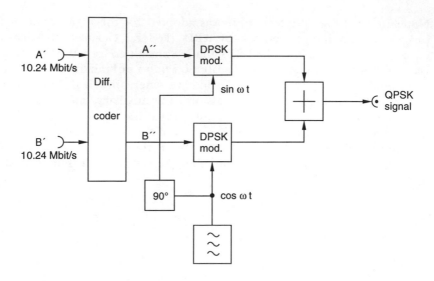

Fig. 2.24 QPSK modulator for DSR

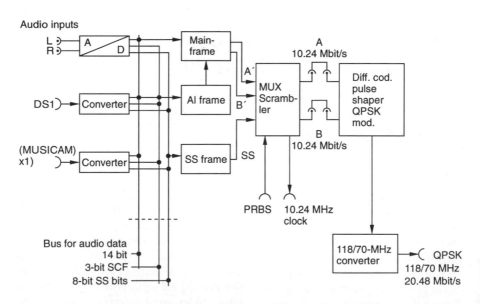

Fig. 2.25 Block diagram of DSR multiplexers and QPSK modulators

x1 (Section 2.4.2)
SS = special service
AI = auxiliary information

SCF = scale factor
PRBS = pseudo random
bit sequence

RF carriers for the two DPSK modulators are phase-shifted by 90°. The DPSK modulators may be regarded as switches for reversing the polarity of the RF signal after differential encoding of the binary data [70].

The DSR multiplexer/modulator may be used for monitoring the bit error rates occurring on transmission links and in transponders. To this end, noise is superimposed on the QPSK signal and the resulting bit error rate (BER) measured as a function of the carrier/noise ratio (C/N). The error protection codes used in the DSR system ensure undisturbed reception to a bit error rate of about 3×10^{-3}. The associated signal/noise ratio can be determined with the aid of the DSR multiplexer/modulator, since modulation can be switched off to determine the C/N ratio (CW mode). For measuring the BER, a random bit sequence derived from a bit error measuring unit has to be fed via the PRBS (pseudo random bit sequence) interface. The 10.24 MHz clock output is used for synchronising the applied bit sequence [61].

2.3.2.4 DSR pulse frame
The two main frames A and B are composed of 320 bits each with a repetition frequency of 32 kHz. The overall bit rate of each frame is 10.24 Mbit/s. Within a frame, a 10-bit sync. word is followed by a special service bit and four 77-bit blocks (Fig. 2.26).

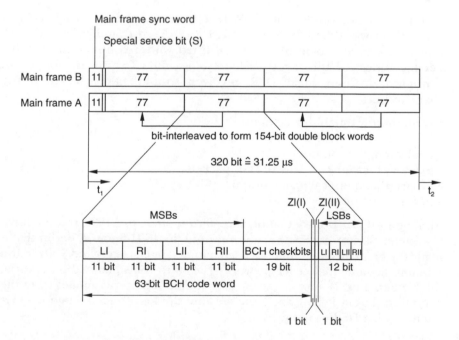

Fig. 2.26 DSR pulse frame (I, II = stereo channel numbers)

The special service bit transmitted in main frame A forms the special service superframe consisting of a total of 512 successive bits (eight special service frames), in which all programme identifications (mono/stereo, speech/music, programme type) of all 16 stereo programs are transmitted every 16 ms. The four 77-bit blocks are interleaved in pairs bit by bit to distribute bit errors over several programs. Within the 77-bit blocks, the first 11 bits of the 14-bit codewords of two stereo programmes are followed by 19 protective bits formed according to the BCH method (block code by Bose, Chaudhuri and Hocquenghem). This code allows up to three errors within the protected 44 bits to be corrected and in addition one error to be concealed. The BCH 63/44 code is followed by two AI (auxiliary information) bits and the remaining 4×3 bits of the codewords [60, 61].

An AI frame with the scale factors for each stereo programme and programme information are transmitted in the auxiliary information bits.

2.3.2.5 DSR receivers

Prices of DSR consumer receivers have dropped below £220 since mid-1993 with about 40 000 units having been sold in 1991 and a further 100 000 in 1992. About 80% of homes receive DSR via cable. Due to its high emitted power, TV-SAT2 is perfect to extend DSR to include small portable DSR radios with integrated flat antennas with an edge length of, say, a mere 15 cm. These ideas will undoubtedly take a few more years to become reality [67].

Opinions are divided on DSR reception via satellite, since DSR is currently provided both by KOPERNIKUS (23.5° east) and by TV-SAT (19° west). As a result, some receivers are fitted with a continuously tunable satellite tuner for the range 950 to 1750 [69]. To prepare for future configurations, an ideal DSR receiver should have the following characteristics:

• continuously tunable cable input
• continuously tunable satellite input
• field strength indicator
• memory locations for 5 to 10 DSR packets
• electrical and optical digital output
• headphones output.

Although official DSR operation was started on at the 1989 International Consumer Electronics Show, it was not until the 1991 Show in Berlin that it made the breakthrough. DSR has been adopted by Germany and Switzerland (community antenna feeder network) [68] with trials planned for Austria, the Netherlands, Denmark and possibly China. The programmes from the central studios in Peking are to be fed into the local regional networks via satellite using DSR techniques.

2.3.3 DSR applications

2.3.3.1 Euroradio
Since 1984, the EBU (European Broadcasting Union) has been using leased transponders in the Eutelsat I-F2 satellite for transmitting Eurovision television programmes, and there was an urgent need for exchanging high quality Euroradio programmes, the audio equivalent to Eurovision, via the same transponder. To meet the requirements of EBU for the DS1 technique, coders and decoders had to be combined with the appropriate sound broadcast transmission equipment [72].

The DS1 units were developed for DSR feeder lines and then chosen for the digital TV studio. Because of the good transmission characteristics of these links there was no need for a high level of error protection. The application envisaged by the EBU created a difficult environment for the signal, however. There was the risk of interference from adjacent television channels, and with the use of small receiving antennas the system was required to work with a low signal/noise ratio. This called for much better error protection and a more effective modulation mode by satisfying the demands with a QPSK (quadrature phase shift keying) modulator and the integrated error protection with a code rate of 1/2. The bit rate of 1.024 Mbit/s at the output of the DS1 line transmitter is doubled by the error protection circuit, and the bit stream of 2.048 Mbit/s QPSK is modulated onto a 70 MHz carrier for sending to the earth station.

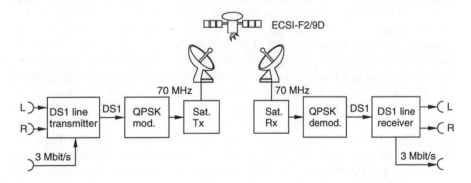

Fig. 2.27 Euroradio system

Various test broadcasts were conducted with an experimental rig (Fig. 2.27) comprising a DS1 line transmitter and a QPSK modulator at the transmitting end, and a DS1 receiver and QPSK demodulator at the receiving end, with 1/2 rate error protection, permitting transmission of high quality sound via a low-power satellite link occupying a bandwidth of only 1 MHz. By modifying the frequency assignment of the transponders it was possible to insert the sound signal between two television signals (Fig. 2.28).

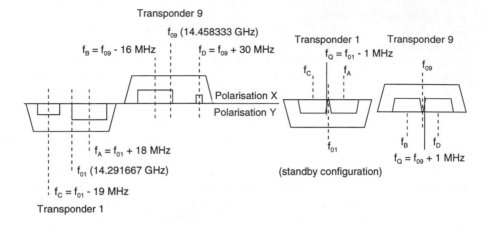

Fig. 2.28 Assignment of transponders in Eutelsat I-F2 satellite (uplink)
(a) Previous frequencies at which sound signal is transmitted together with a TV signal via a 72 MHz transponder
(b) Carrier frequencies for Euroradio; one sound signal and two TV signals are transmitted at a time

Calculations and trial transmissions conducted by working groups of EBU with the support of DBP Telekom, Eutelsat and Telespazio (Italy) have shown that an EIRP (equivalent isotropic radiated power) of 70 dBW and a receiving system with a G/T of 25 to 28 dB/K, will give reliable service covering the majority of Europe. For an EIRP of 70 dBW it is sufficient to have a parabolic dish of 2.5 m diameter combined with a 160 W amplifier. This makes it possible to use temporary erected transportable earth stations, already used by many broadcasters.

The first experimental concert broadcast was from the 1988 International Consumer Electronics Show in Berlin. Besides Berlin, Usingen (DBP) and IRT in Munich, the signals were also received by NOB in the Netherlands and BBC in the UK. For further concert broadcasts from Copenhagen, Munich and Oslo, links-ups were made with TDF (France), DR (Denmark) and RAI (Italy).

At the EBU, the system described here had to compete with other proposals from France and Sweden. It has now been in operation since September 1989.

2.3.3.2 DSQ (digital studio quality)
Following an EBU decision in 1992, the Euroradio system since 1993 onwards is longer based on DS1, but follows the CCITT standards G700 and G800. The baseband signal is coded in line with the CCIR Recommendation 724: 20 bits per sample are reduced to 18 bits with a sampling frequency of

48 kHz using the near-instantaneous companding technique in accordance with the NICAM technique (Section 3.3.3).

This method uses a 1 ms block with eight coding ranges. Sampling interleaving ensures protection against bit errors. The data rate is 2.048 Mbit/s in accordance with CCITT G703-6, and the output signal is HDB3-coded (see DS1, Section 2.3.2). A 67 Kbit/s auxiliary data channel is provided in the transmitted data stream. In addition to the analogue L/R interface, the DSQ interface also has an AES/EBU input interface in line with CCIR Rec. 647 with a data rate of 3.072 Mbit/s.

DSQ from Philips Communication Systems has a 20 kHz bandwidth and a 120 dB dynamic range (DS1: 15 kHz, 96 dB). The system has no significant data reduction to ensure simple postprocessing and avoids the delays that make lip synchronisation and loopback voice monitoring virtually impractical.

2.3.3.3 DIGit super radio

DIGit Super Radio was introduced in Switzerland in 1991. During 1992, over 60 cable network operators ordered DIGit programmes and from mid-1992 about 0.7 million subscribers could receive DIGit programmes. A continuously tunable DSR receiver connected to the digitally compatible cable outlet is all that the digital radio listener needs to install. Unlike the DBP TELEKOM, which simultaneously distributes programmes via the communications satellite DFS1-KOPERNIKUS and the direct broadcasting satellite TV-SAT2, feeding them to the cable headends, the digital programmes in Switzerland are distributed via a terrestrial microwave link direct to the local antenna feeder network [73].

Other European countries such as France, the UK, Austria, the Netherlands and Italy are also interested in the DSR system, with a view to transmitting DSR programmes via satellite cable networks.

2.4 Audio baseband coding

In the field of both professional and consumer equipment the development of audio technology is being led by digital techniques. High performance signal processors and user-oriented ASICs enable complex algorithms to be implemented as, for example, required for source coding. The reason for data compression in source coding is that, for instance, 16-bit linearly coded signals would require too high a transmission bandwidth and too much storage capacity in the equipment. This would increase the cost of satellites fees, lines and storage elements. Another important point is the optimum use of the frequency resource. In particular, with a view to using terrestrial transmitters for digital mobile reception of sound programmes it is necessary to reduce the data rate of the sound signal by approximately a factor of eight compared to the CD. Using the findings of psychoacoustics on human hearing, high quality audio signals in CD, studio or concert quality can be reduced to values below 200 Kbit/s per stereo signal. Relative to a sampling frequency of 48 kHz, this means a quantisation of the audio signal of only about 2 bits per sample. Despite this high data reduction, there are hardly any perceptible impairments.

2.4.1 Psychoacoustics

Audio source coding uses the findings of psychoacoustics [74, 75]. Data reduction of the source signal is based on the redundancy (unnecessary information) and irrelevance (imperceptible information) contained in the part of the audio signal which the human ear cannot perceive. The irrelevant signal components do not contribute to the definition of a characteristic, i.e. to the identification of the sound event, to the sound quality or the localisation. They have no significant influence on human hearing and subsequent information processing. Irrelevance reduction means that, firstly, irrelevant signal components will not be coded and, hence, are not transmitted and, secondly, distortion resulting from low quantisation (quantising noise N_q) will be tolerated provided that such distortion is not perceived by the human ear due to masking.

$$N_q = n \times 6.02 + 1.76 \text{ [dB]}, \quad n = \text{linear quantisation, e.g. 16 bits}$$

The effect known as the resting threshold (threshold of hearing) represents the perceptible sound level as a function of frequency (Fig. 2.30, Fletcher/Munson curve).

A second effect is masking. Masking describes a psychoacoustic characteristic of the human ear by which low-level audio signals are masked by neighbouring high-level audio signals. A special aspect of masking is that a soft signal is masked not only at the instance of a loud signal. Masking also takes place with the soft signal present after or even before the loud tone (the masker). The latter effect is described as premasking (Fig. 2.29). Depending on the temporal characteristic of masking, distinction is made between pre-, simultaneous and postmasking of an audio signal. Simultaneous masking has the greatest efficiency in respect of data reduction.

The masking effect, which is frequency- and level-dependent, can be substantiated by listening tests. With a constant masker, the test tone just masked is recorded in frequency and level. The resulting frequency-dependent level characteristic is called the masking threshold. The masking effect also applies to multiple tones. It is therefore the task of data reduction to separate the masked signal component from the maskers that are significant to the human ear. Only the components perceptible to the human ear will be coded. The signals above the masking threshold are however coded with low quantisation so that the quantising noise produced remains below the masking threshold.

Quantisation is determined by the ratio of the signal amplitude to the

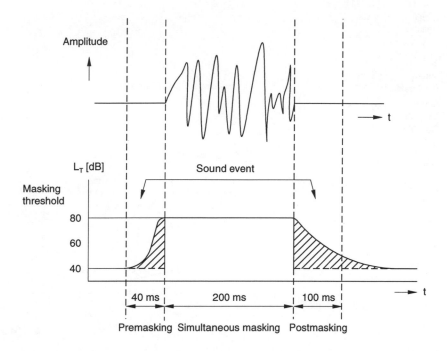

Fig. 2.29 Pre- and postmasking of an abrupt sound event as a function of time [76]

minimum masking threshold SMR (signal/mask ratio).

Investigations into masking thresholds have shown that the human ear divides the masking signal into frequency bands (Fig. 2.30). These bands have a constant absolute bandwidth below 500 Hz and a constant relative bandwidth above 500 Hz. This frequency effect on the ear has to be taken into account when dividing the audio baseband signal into subbands.

Depending on the type of analysis and synthesis of the audio signal a distinction is made between transform coding and subband coding.

The transform coding method transforms a time window of the input signal into the frequency domain, using an FFT (fast Fourier transform) or a DCT (discrete cosine transform). The amplitude and phase values in the frequency domain are quantised, coded and transmitted by psychoacoustic effects. The decoder expands the frequency values and transforms them back into the time domain.

The subband coding method divides the audio signal into subbands, using QMF (quadrature mirror filters), digital wave filters or polyphase filters [77]. The individual subbands are sampled and subjected to data compression. The data compression is dependent on the masking threshold which is calculated at certain time intervals. In the receiver, the data are expanded by an inverse filter bank to retrieve the original audio signal.

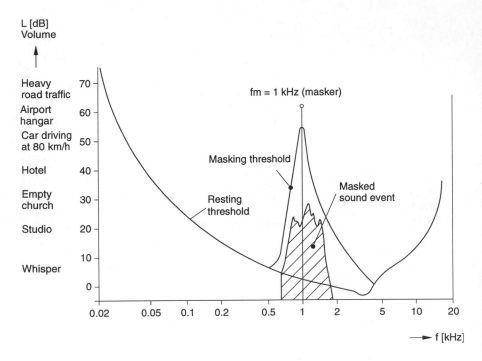

Fig. 2.30 Audio resting threshold and masking threshold of a pure 1 kHz audio signal

The advantage of dividing the baseband signal into many frequency subbands is that both transform and subband coding bit errors are confined to narrow frequency bands and are therefore less disturbing than PCM systems, where the signal is digitised and transmitted in broadband form [78].

2.4.2 MUSICAM

In 1988, IRT developed baseband coding in line with MASCAM (masking-pattern adapted subband coding and multiplexing) [82]. In co-operation with CCETT (Centre Commun d'Etudes de Télédiffusion et Télécommunications) in France and Philips in the Netherlands, IRT further developed the technique and created the MUSICAM method. Beyond its use in digital audio broadcasting [83], this method has gained worldwide acceptance as a universal baseband coding method for high quality audio signals based on subband coding and using the redundancy and irrelevance contained in the music signals for effective data compression.

The MUSICAM method uses 32 subbands. The subbands are obtained by means of a polyphase filter bank and each has a bandwidth of 750 Hz. The sampling frequency of each subband is 1/32 of the 48 kHz output frequency, i.e. 1.5 kHz. Blocks are generated from the samples of the subbands (12 consecutive samples per block). A scale factor is derived from each block by determining the peak level and coded with 6 bits. The scale factor, which is the maximum level within a subband block, can be represented by 6 bits, i.e. 64 volume classes. Each volume class corresponds to a 2 dB range, yielding a total dynamic range of about 120 dB. The scale factor is used for block companding, which means that the point of quantisation is shifted as a function of the level.

The masking threshold for each subband is calculated according to the psychoacoustic model, taking into account both the masking and the quantising noise from the resting threshold which reduces quantisation.

The filter bank in the MUSICAM coder (Fig. 2.31) is a polyphase filter which is less complex and features a lower time delay than the better known quadrature mirror filter [77]. The overall delay of MUSICAM codecs caused by filtering in the coder and inverse filtering in the decoder is 11.6 ms at a sampling frequency of 48 kHz.

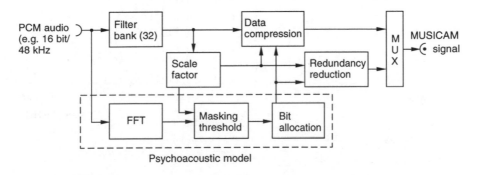

Fig. 2.31 Block diagram of MUSICAM coder

The constant 750 Hz bandwidth of the MUSICAM subbands is not an ideal frequency for the masking effect of the human ear below 500 Hz. With an optimum subdivision, the first subband from 0 to 750 Hz would contain about seven frequency groups. To avoid an unnecessarily fine quantisation in this range or a too coarse one that would cause audible noise, an FFT with a window length of $\Delta t = 1024/f_s$ ($\Delta t = 21.33$ ms at $f_s = 48$ kHz) is carried out in parallel to the audio signal filtering. This FFT is used to compensate for the inaccuracy of spectral analysis caused by the filter bank in the lower frequency range. The FFT allows accurate determination of the masking thresholds for clear bit allocation to the subbands. FFT is not switched into the signal path and only employed in the coder. This complex configuration is not required in the decoder, an important aspect in the development of consumer products.

Fig. 2.32 shows the spectrum and the equivalent masking threshold of an audio signal in a short time window. There is underlying quantising noise resulting from adaptive coding. Subbands below the masking threshold are not necessary for the human ear and are therefore not transmitted.

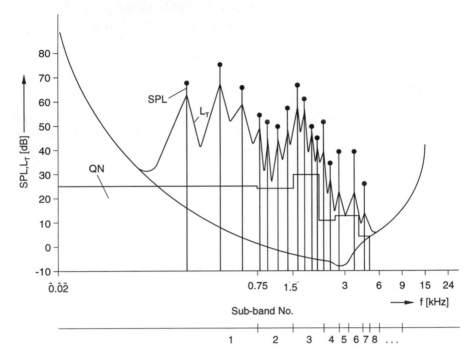

Fig. 2.32 Masking threshold and quantising noise of a typical vowel [77]
SPL = amplitude spectrum (typ. vowel) with 16 subtones L_T = masking threshold
Q_N = quantising noise

The dynamic bit allocation, i.e. the actual calculation of the masking threshold and derived quantisation of the subband signals, is performed every 24 ms. This interval corresponds to the mean dwell time of speech and music signals [78, 79]. Since the audio signal is naturally subjected to continuous change, the filtered audio data are variably quantised according to the bit allocation, resulting in a dynamic bit rate. Apart from the dynamic bit rate for the transmission of the coded samples, the multiplex signal to be transmitted contains the additional information required for the decoder, i.e. scale factors and information about the the bit allocation.

The bit rate on the transmission link must be constant, so buffer memory may be used to stabilise the dynamic bit rate at a constant value. This buffer will, however, cause a delay which has to be kept minimal for many applications. Therefore, blockwise stabilisation of the dynamic bit rate at a constant value, e.g. 96 Kbit/s, is increasingly being adopted. At higher bit rates, e.g. 128 Kbit/s per mono signal, a dynamic bit rate margin is usually available. This bit rate margin may be used, for instance, to quantise the subbands at a higher level than required and to obtain a higher mask/noise ratio. The variable bit rate margin may be used to provide the signal with high error protection in pauses and low-volume passages which are particularly prone to bit errors, thus ensuring high quality transmission. A third possibility of utilising the dynamic bit rate margin is the transmission of additional noncritical programme-related information, similar to that described for the RDS system (Section 2.2).

MUSICAM covers bit rates of 32, 64, 96, 128 and 192 Kbit/s for the mono signal (Fig. 2.33). A bit rate of 64 Kbit/s yields for many signals a subjective quality comparable to the compact disc. This bit rate would appear suitable for use in ISDN (integrated services digital network), such as outside broadcast links.

Bit rates of 96 and 128 Kbit/s per mono signal equate to CD quality. These bit rates are used for DAB (digital audio broadcasting) sound transmission and HDTV (high-definition television) with several sound channels providing storage of high-quality audio signals. The bit rate of 192 Kbit/s provides studio quality, as the coded sound signals feature a quality margin. Typical postprocessing in the sound studio using effects, cutting, filtering and compression techniques does not cause any noticeable degradation (Fig. 2.33).

The MUSICAM coder provides the possibility of selecting sampling frequencies of 32, 44.1 and 48 kHz enabling sound signals to be directly coded to the current standards such as AES/EBU.

The MUSICAM decoder contains the reciprocal functions of the coder, i.e. demultiplexing, decoding, inverse data compression and inverse filter

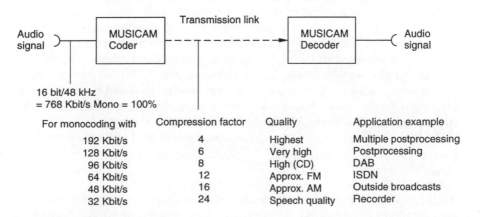

For monocoding with	Compression factor	Quality	Application example
192 Kbit/s	4	Highest	Multiple postprocessing
128 Kbit/s	6	Very high	Postprocessing
96 Kbit/s	8	High (CD)	DAB
64 Kbit/s	12	Approx. FM	ISDN
48 Kbit/s	16	Approx. AM	Outside broadcasts
32 Kbit/s	24	Speech quality	Recorder

Fig. 2.33 MUSICAM data compression

bank (Fig. 2.34). In the decoder, up to three scale factors and the bit
allocation are derived from the multiplex signal for one time block of 3×12
samples for each subband. The subbands are decoded from these signals and
restored to their original, typically 16-bit format. The filter bank is the most
intensive processing part of the decoder in retrieving the original sound
signal from the subbands. Consideration has been given to the MUSICAM
specification in simplifying the design of the decoder to minimise CPU
(central processing unit) processing time. IRT has designed a decoder
around the Motorola DSP56001 24-bit width signal processor or the AT&T
DSP32C 32-bit width signal processor [77, 78].

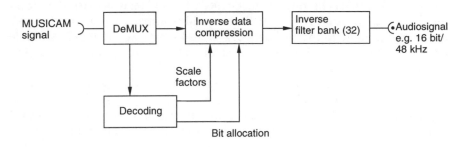

Fig. 2.34 Block diagram of MUSICAM decoder

   The MUSICAM codec implemented by IRT yields an overall delay of
45 ms at a sampling frequency of 48 kHz with the coder delay amounting to
35 ms. This overall delay is tolerable for all applications envisaged so far. The
use of a buffer memory is recommended for a data rate of 64 Kbit/s to
reduce peak values in the dynamic bit rate characteristic. Considering the
delay arising from the buffer, a total delay of 93 ms is obtained at a bit rate
of 64 Kbit/s. The bit error sensitivity of MUSICAM-coded signals can be said
to be very good. Investigations have shown that with a bit error rate of 10^{-3}
the subjective quality of the audio signal is practically unimpaired [76].
   One of the aspects in the future development of MUSICAM is the so-
called joint stereo, with redundant and irrelevant components in the
left/right stereo signal being further reduced. Other considerations also
include the cinema surround sound technique with the transmission of five
channels: left, centre, right and surround-left and surround-right as
background.

2.4.3 ASPEC
Audio baseband coding in line with ASPEC (adaptive spectral perceptual
entropy coding) is a joint development with the Fraunhofer Institute of
Erlangen, University of Erlangen, Deutsche Thomson-Brandt, CNET

Research Institute of the French PTT and the laboratories of the US company AT&T. The block diagram of the ASPEC coder (Fig. 2.35) comprises the following functional blocks.

The input buffer storing the digitised audio signal in one block is followed by data transformation into the frequency domain. The transformation is a modified MDCT (discrete cosine transformation) which compensates aliasing effects in the time domain. The MDCT is equivalent to a 512-band filter bank [78, 81].

Quantisation and coding of the transform coefficients follows below, taking into account the masking threshold. The Huffmann code is used for coding, the shortest codewords being assigned to the most frequently occurring spectral values in the sense of an entropy. Huffmann coding is based on code tables with individual codewords which are permanently stored in the coder and decoder, and forms an essential part of the overall data reduction.

Quantisation and entropy coding is controlled by a parallel block, in

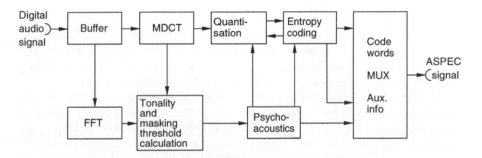

Fig. 2.35 Block diagram of ASPEC coder [79]

which the masking thresholds are estimated on the basis of the transform coefficients of the MDCT. For accurate calculation of the masking thresholds, an FFT is simultaneously performed for ASPEC, similarly to the MUSICAM algorithm. In contrast to MUSICAM, FFT is not used for its accurate frequency resolution, but to distinguish tonal components from noise components with the aid of pointer prediction. The noise components must be quantised at a lower level during coding, since they are naturally less affected by the quantising noise. The tonal components have to be quantised to ensure that the quantising noise is below the masking thresholds.

The fourth functional block of the ASPEC coder contains error protection to improve the transmission reliability of the transform coefficients. The proper codewords and the additional information are combined in a multiplexer to output as a bitstream.

The ASPEC decoder mainly contains the reciprocal functions of the coder: in the first block, errors are corrected and the Huffmann-coded data decoded in a simple finite look-up table (Fig. 2.36). Scaling and windowing performed during quantisation in the coder are then revoked and the MDCT spectrum reconstructed. In the subsequent inverse MDCT the pulse-coded data stream originally applied to the ASPEC coder is retrieved [79].

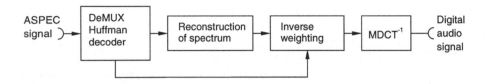

Fig. 2.36 Block diagram of ASPEC decoder [79]

2.4.4 ISO Standardisation

MUSICAM and ASPEC are complex data reduction methods employed in audio source coding and based on complex developments and basic research in psychoacoustics. An objective assessment of the two methods regarding their suitability to becoming a world standard is equally difficult.

In addition to audio quality being the main parameter, other criteria such as resistance to bit error or low complexity of decoder (low price) were also considered in the evaluation. The ISO WG 11 Audio Group [80] (International Standardization Organization Working Group) not only tested and evaluated MUSICAM and ASPEC but also other alternative source coding methods to defined test parameters and associated weighting criteria. In addition to objective criteria, i.e. measurable parameters, subjective quality criteria were also integrated in the assessment. To sum up briefly, the results showed that the advantages for MUSICAM were in the objective tests, while ASPEC had advantages in the subjective listening tests, in particular at low bit rates (64 Kbit/s/mono). This is why the ISO Audio Group decided to combine the two methods to capitalise on the advantages of both MUSICAM and ASPEC. The combination of the modules allows different data rates and stages of complexity to be implemented (multilayer structure).

The outcome was a standard defining three layers (Table 2.6):

• Layer I is a MUSICAM subband coding of simple structure intended for a relatively high bit rate.
• Layer II is designed for medium bit rates.
• Layer III includes elements of transform coding for higher frequency resolution, dynamic windowing and entropy coding from the ASPEC method. Layer III is intended for the lower bit rates.

Layer I	MUSICAM, simplified 192 kbit/s mono low complexity	Consumer applications Storage media e.g. PASC in DDC
Layer II	MUSICAM 128 kbit/s mono medium complexity	DAB Professional applications
Layer III	Subband/transformation code MUSICAM/ASPEC 64 kbit/s mono high complexity	ISDN Telecommunications

Operating modes: mono, dual channel (multilingual, stereo, joint stereo coding)

Table 2.6 ISO layers in specification 11172-3

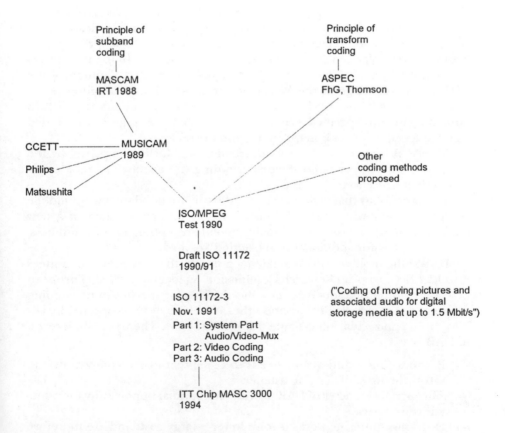

Table 2.7 History of ISO Standard 11173

Layer-specific information is added to the data stream for the automatic adjustment of the decoder. All layers use a fixed coding algorithm allowing a simple design of the decoder [78]. This standardisation includes both the transmitted data format and the decoder. There is a free choice regarding the type of coding of the source coder, although an accurate psychoacoustic evaluation requires a coder of high complexity.

ITT developed the MUSICAM coder and decoder: MASC3000 (multipurpose audio signal compander) with the main emphasis on the MUSICAM decoder itself, designed with coder and decoder on the same chip. The MASC3000 chip is implemented with a RISC processor that is used to approximately 40%. The codec was initially built for layers I and II. For layer III, only the decoder function will be implemented. The MASC3000 chips have been available from mid-1993 (Table 2.7) [82, 84].

2.5 DAB (terrestrial digital audio broadcasting)

VHF FM sound broadcasting no longer meets the advanced requirements of today's, let alone future, technology. When VHF FM sound broadcasting was planned, stationary reception was assumed to be a basic prerequisite. Networks were planned for a receiving antenna height of 10 m with a three-element directional antenna for stereo signals. The later planning of the VHF frequency range at the World Administrative Radio Conferences in Geneva in 1982 and 1984 meant that a directional antenna with 12 dB antenna gain was specified explicitly for the VHF coverage but excluded suitable specifications for in-car radios and portables.

Mobile reception of VHF sound broadcasts is further impaired by the increasing number of radio stations crowding the airwaves causing cross-channel interference.

The probability that digital satellite radio will not be adaptable for mobile reception either was sufficient motivation for the development of a new method of achieving the best in quality, mobile reception, user friendliness, and capabilities for additional (value-added) services.

DAB at the moment is a great talking point. In the European committees Eureka 147 (European Research Commission Agency) and ETSI (European Telecom Standardisation Institute) the drafting of specifications is going ahead at full speed. In Europe and other countries in the world field trials have been conducted using single transmitters [85]. The immense interest in DAB is:

— Broadcasting sound quality needs to equal that of the compact disc to satisfy the most discerning listener.
— The digital character of DAB provides an optimal opportunity for value-added services.
— DAB transmitter networks involve lower energy costs and are therefore attractive because they are 'green'.

The keywords of DAB are: MUSICAM, COFDM and SFN (single frequency network).

- MUSICAM is a method for baseband coding of audio signals. By utilising psychoacoustic effects, the bit rate of, for instance, a 16-bit linearly coded mono signal at 48 kHz sampling frequency can be reduced to 96 Kbit/s, i.e. by a factor of eight.
- COFDM mainly solves the problem of terrestrial multipath reception. Echo signals are used to add to the direct signals. The technique is based on allocating the data stream to many carriers, with the individual carrier being DQPSK modulated, an orthogonal carrier arrangement, introduction of a guard interval for the utilisation of multipath signals and interleaving of the programme signals in the time and frequency domain. COFDM (coded OFDM) is a specific coding method for DAB patented by the French CCETT and a form of the known OFDM method. In the following, the 'C' will be regarded as a part of coding while 'OFDM' represents the multicarrier modulation.
- The third DAB keyword — SFN — means that the transmitters of a DAB network operate at exactly the same frequency, and furthermore the radiated OFDM signals are synchronous in clock and frame within a tolerance of 5% of the guard interval, in other words, exactly the same bits are radiated at any given time.

2.5.1 DAB system technique

2.5.1.1 Overview

The individual functional blocks of a DAB transmitter network are shown in Fig. 2.37. The programme distribution to the terrestrial network transmitters, for example, via satellite is not covered here but in Section 4.4.2. The first suitable programme distribution points are after baseband coding, that is after the MUSICAM coders. Otherwise another possibility is after channel coding and multiplexing.

In a field trial using a 30/20 GHz satellite link the analogue interface was inserted after the OFDM modulation (Section 2.5.7). This solution will not, however, be considered for DAB operation when it commences in 1995.

The MUSICAM coding shown in Fig. 2.37 for reasons of bit economy is designed to be close to the audio source. It is followed by channel coding and multiplexing, e.g. of six stereo programmes with 96 Kbit/s per mono signal. In conjunction with an organisation and data channel, synchronisation symbols as well as an error protection, a gross bit rate of 2.4 Mbit/s is obtained.

To enhance flexibility, a data service may in future be inserted to replace the sound signal. Transmission of still pictures, detailed traffic information or radio paging services are also being considered.

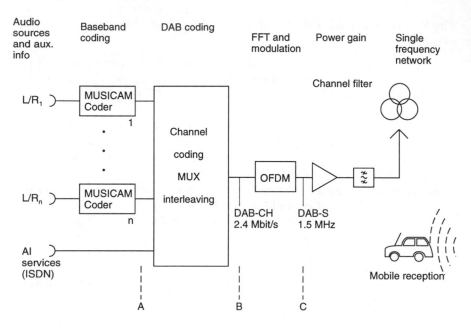

Fig. 2.37 Block diagram of DAB transmission (the interfaces A, B, C) are alternative programme distribution points to the transmitter sites)

Channel coding has the task of providing the source coding bits with optimum error protection for the subsequent channel. Channel coding should enable error detection and — to a certain extent — error correction. One possible solution is the use of block codes, a block of net bits being assigned a certain number of bits for error protection. Convolutional codes, in which the individual net bits are continually assigned bits for error protection, are another possible solution.

DAB uses the average code rate of one-half, which means that each net bit is assigned an additional bit for error detection and error correction. The error protection bits may be unequally distributed according to the significance of the bits transmitted — UEP (unequal error protection). Since with terrestrial multipath propagation uncorrectable burst errors would be caused by signal fading, the transmitted bits are interleaved in the channel coding. By this method, burst errors are split up into individual bit errors which can be corrected by the error protection of the channel coding (Fig. 2.38).

Channel coding is followed by OFDM in the signal processing path. The principle of OFDM in using many carriers, each being DQPSK modulated. Modulation and demodulation of such a frequency multiplex signal (multicarrier method) is performed effectively using digital signal processing and IFFT (inverse fast Fourier transform) as well as FFT.

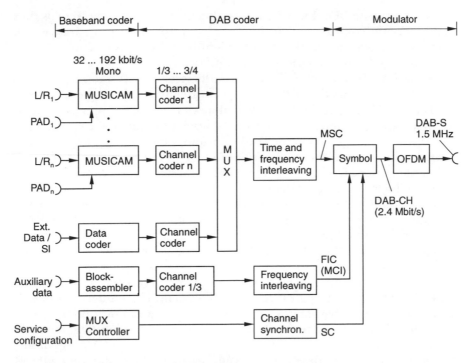

Fig. 2.38 Block diagram of DAB generator to EU147 [92]

In the Rayleigh channel the subcarriers may become distorted as a result of signal fading. A possible correction measure in the receiver is the use of a Viterbi decoder and soft-decision technique. Burst errors should, however, be avoided for the operation of the Viterbi decoder. This led to the method of assigning consecutive bits in the data stream to subcarriers that are widely spaced apart in the transmission channel (frequency interleaving).

An orthogonal carrier arrangement is achieved if the spacing between the individual carriers of the multicarrier signal is equal to the frequency that corresponds to the digital modulation step size. The spectra with the zero crossings of the $\sin x/x$ function are then overlapped and clearly separated in the receiver. Each individual carrier of the multicarrier system is DQPSK (differential quadrature PSK) modulated [94].

In the multipath channel there are delays which may cause interference in a monocarrier system. In the multicarrier system, where the step frequency of the modulation signal is much lower because of the number of carriers, a guard interval can be additionally inserted which must be greater than the expected maximum delay. In the DAB system, the guard interval is one-quarter of the modulation step size. The mobile receiver ignores the incoming main signal and the multipath signals during the guard interval and only evaluates the time of the modulation steps. The echo signals thus make a useful contribution to the signal [86, 87]. The price for perfect multipath transmission is the loss of one-fifth of the transmission time and data capacity due to the inserted guard interval.

Regarding the amplifier element of a terrestrial DAB transmitter it has to be considered that the ODFM signal is made up of 1536 single carriers in total (mode 1). Adding up the individual modulated carrier signals yields an RF signal with an amplitude characteristic similar to white noise. This means that the power amplifiers in the DAB transmitter must feature an extremely wide dynamic range. The average power of an OFDM signal measured with a thermal power meter is much lower than the stochastic peak power [96]. The amplifier must be designed to handle such peaks since nonlinear limiting would cause distortion within and outside the transmission band. This means that only high-linearity analogue amplifiers are suitable for DAB. Although the transmitted average power is only about one-tenth of that of an FM transmitter, the DAB amplifier has to be designed for ten times the average rated power.

DAB uses single-frequency networks to cover large areas. Conventional VHF FM transmitter networks are made up of many transmitters in the area to be covered, each having its own frequency, but transmitters broadcasting on identical frequencies must be out of range of each other. Theoretically, 31 channels would be required to cover large areas by one VHF programme.

DAB transmitters operate at the same frequency with the same OFDM signals. Each transmitter sends the same bit at the same time. The single-frequency capability achieved through the introduction of the guard interval is an optimum solution for uniform coverage [88, 90].

2.5.1.2 Channel model

The channel model illustrated in Fig. 2.39 is assumed for the transmission of DAB signals. The DAB transmitters A and B operate at the same frequency with synchronous clock and bits. The signal from transmitter A is routed to the receiver via various paths a_0 to a_n with the assigned delay times t_0–t_n. Each path is attenuated by h_0 to h_n and modulated. Due to the single-frequency mode, signals from one or several neighbouring transmitters, in our example from transmitter B, may also arrive at the receiver. With their delay time t_{n+1} they appear as an active echo with the path attenuation h_{n+1} so that the receiver cannot distinguish them from the signals of transmitter A. Depending on the phase of the arriving signals, the carriers may be additive or subtractive (fading). A cancellation will, however, only occur in very narrow frequency bands. The larger the bandwidth of the transmission channel, the lower the fading. With the bandwidths suitable for DAB, fading is typically 10 dB for a flat terrain. The field strength present at the receiving site during the useful symbol period is decisive for the reception quality [89].

The basic requirement for unimpaired broadcast reception is a strong field strength at the receiver at all locations and times. Due to the single-

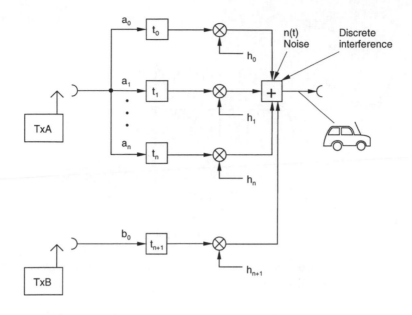

Fig. 2.39 DAB channel model

frequency mode, the sum of all subwaves from neighbouring transmitters contributes to the useful field strength as long as the guard interval is not exceeded. In unfavourable terrain, delays up to 150 μs, corresponding to 50 km, have been measured. Because of the guard interval of 250 μs duration (mode 1), the difference between the nearest transmitter and other transmitters should not exceed 90 km (corresponding to 1.2 × guard interval) without an adverse effect on the reception. Interfering signals should have a minimum spacing of about 10 dB from the direct signal to ensure interference free reception (Fig. 2.40).

The following, mutually independent effects have to be considered in defining the DAB parameters:

- Due to the time delay of the channel impulse response, consecutive symbols affect each other (intersymbol interference), if the data rate becomes too high and consequently the symbol period too short, cancellations may occur (frequency selectivity).
- The channel transfer characteristic changes due to the different sites of the mobile receiver. Phase assessment by the receiver (DQPSK modulation of carriers) deteriorates if the data rate is too low and the symbol period too long (Doppler effect, time selective fading) [97].

To achieve a satisfactory solution, a tradeoff is made between the number of carriers, the transmission channel bandwidth, and the derived data rate per

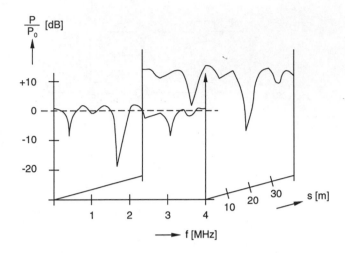

Fig. 2.40 Typical field strength characteristic as a function of frequency and path
(P_0 = mean field strength)

carrier. Since the channel impulse response extends over several microseconds, the effect of frequency selectivity is limited to well below 1 MHz. Therefore, a great number of carriers are chosen which are modulated with a low bit rate. The frequency selectivity becomes effective only for a fraction of the total data stream. Moreover, the guard interval with a longer duration than the channel impulse response ensures that the useful symbol period is kept free from intersymbol interference. Although OFDM solves the problem of channel selectivity, it does not suppress local fading. The main waves are scattered in the direct vicinity of the vehicle. The power spectrum of the nearby reflections depends on the direction and the speed of the vehicle (local Rayleigh fading). The fading produces block errors which are noncorrectable; therefore, time and frequency interleaving is introduced; thus burst errors become single errors which are correctable.

Fig. 2.41 illustrates the channel in the time and frequency domain. The small areas mark zones in which the channel can be considered as locally invariant. The large areas designate the distances which should have correspondent small areas in order to be statistically independent [95]. The frequency domain of the small areas is determined by the spread of the channel impulse response and the time domain by the maximum Doppler frequency. The efficiency of OFDM modulation is thus determined by the subdivision of the time-frequency domain into small areas in which the transmission channel can be assumed to be invariant [95].

2.5.1.3 Doppler effect
The magnitude and phase of the channel transfer function must be

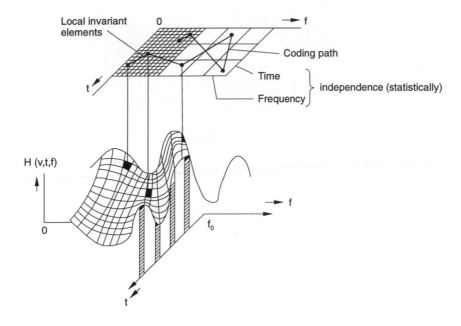

Fig. 2.41 Channel response (field strength) in time (t) and frequency (f) domain with mobile reception [95].

The upper signal is the generated signal and the lower signal is the received signal in a Rayleigh channel

approximately constant for the total symbol period. The period, in which the channel is invariant in time, depends on the Doppler frequency (Table 2.8).

A different approach can be derived from DQPSK modulation (differential PSK) which allows the phase modulator to decide between two defined phase positions of ±45°. Through the frequency shift during the symbol period, the Doppler effect produces a phase shift which reduces the decision margin between the two phase positions for consecutive symbols [97]. The phase shift $\Delta\varphi$ is given by:

$$\Delta\varphi = T_s \cdot f_d$$
$$\Delta\varphi \ll \pm 45°$$

For the Doppler effect, the critical case is illustrated in Fig. 2.42, and can be considered to have merely theoretical significance. The car with the receiver is driving at high speed and the receiving antenna sees two signals. The first arrives in front of the car and produces a positive Doppler frequency while the second with almost the same amplitude arrives from the rear of the car and produces a negative Doppler frequency [95].

Basic equatations

$$f_d = \frac{f_0 \cdot v}{c}$$

f_d Doppler frequency
f_0 Carrier frequency
v Vehicle speed
c Speed of light

$$T_s = t_s + \Delta$$

T_s total symbol period
t_s useful symbol period

$$\Delta' < \Delta \cdot 1{,}2$$

Δ Guard interval
Δ' permissible delay (factor 1,2 due to attenuation echos)

$$d_x < \Delta' \cdot c$$

d_x Transmitter distance for SFN

Condition : $T_s \ll \dfrac{1}{f_d}$ Assumption: $T_s < \dfrac{1}{25\, f_d}$

Example 1:   Example 2:

Mode 1, f_0 = 230 MHz v = 150 km/h f_d = 31.9 Hz   Mode 3, f_0 = 1.5 GHz, v = 150 km/h, f_d = 208 Hz

$T_s < 1.254$ ms
$\Delta < 251$ µs
$\Delta' < 301$ µs
$d_x < 90$ km

$T_s < 192$ µs
$\Delta < 38.4$ µs
$\Delta' < 46$ µs
$d_x < 13.8$ km

Table 2.8 Boundary conditions of Doppler effect

Fig. 2.42 Critical case for the Doppler effect

2.5.2 DAB generation and transmission

2.5.2.1 Multiplex sound signal

Fig. 2.43 shows the configuration of the transmission frame with the multiplex sound signal for six stereo signals, reference symbol, run-in symbol and the null character.

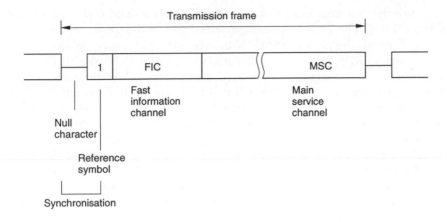

Fig. 2.43 DAB transmission frame

To allow for maximum use of the DAB system, it is advisable not to rigidly fix the overall data volume to the different services. The network operator should be able to adapt the configuration of his data channels to new applications. A possible approach is to introduce different sound signal data rates provided by MUSICAM in the multiplex signal. By this method, multitone channels could be configured with lower quality. After deduction of the reference symbol from the net capacity of typically 1.2 Mbit/s (0.8 to 1.7 Mbit/s), the useful data rate is sufficient for a larger number of stereo signals.

A null character, at which the DAB transmitter does not radiate energy, is inserted in the multiplex signal. The receiver uses this energy gap for a stable synchronisation. Moreover, the interference status of the transmission channel can be analysed during this transmission-free time to localise discrete noise signals.

During the reference symbol, the DAB transmitter supplies a fixed OFDM signal which is used to start differential demodulation. It can also be used to calculate the channel impulse response and to optimise the period of evaluation.

The DAB specification distinguishes between two channels in the multiplex: the MSC (main service channel), approximately 2.3 Mbit/s, and the FIC (fast information channel). The main service channel contains the audio services along with various data services: PAD (programme associated data), MCI (multiplex configuration information) and SI (service information). Code rates of one-third to three-quarters (high to low protection) can be selected for the individual services. The fast information channel contains information about the current or future contents of the main service channel (table of contents, data rates, degree of error protection). The information of this channel is repeated frequently to enhance the transmission reliability. The FIC has a fixed·code rate of one-third [91,92]. The channel is synchronised with 2 Kbit/s via the multiplex controller with the aid of the SC (synchronisation channel) signal.

Similar to RDS, auxiliary information is also transmitted with DAB and subdivided into programme-associated and programme-independent data. The programme-associated data is similar to that of the RDS system. The PAD (programme-associated data) is integrated into the MUSICAM data stream, since MUSICAM reserves 2 Kbit/s for PAD in each mono signal irrespective of the selected bit rate and mode (mono, stereo, dual-channel).

The capacity for programme-independent data can be gained through a suitable configuration of the main service channel, while also consideration is being given to using various RDS in-house applications such as TMC (traffic message channel), EWS (emergency warning system) and other value-added services [90].

2.5.2.2 Channel coding

DAB channel coding uses a convolutional code which allows unequal error protection of the net bits provided by baseband coding. This unequal error protection further improves the transmission quality since, for instance, an MSB (most significant bit) or a scale factor for a MUSICAM subband can be protected at a higher degree than the LSB (least significant bit) of a word. The check data and control information for the decoder should be given the highest degree of protection. A special case of the fixed channel code rate of one-half, that is, one protection bit per net bit, is the transmission of unknown or homogeneous data, as, for example, in the case of texts. The change from variable to fixed code rates and vice versa with the structure of the variable code rates has to be transmitted to the receiver with maximum reliability.

The variable code rates are implemented by puncturing, i.e. suppression of certain error protection bits. This is done according to a defined algorithm, the puncturing scheme, known to the receiver for decoding. There are considerations to use different puncturing schemes to provide varying degrees of protection required by source coding (adaptive punctured codes) [86], and, of course, the decoder has to be informed of the puncturing scheme used.

2.5.2.3 Principle of OFDM

The method chosen by EUREKA 147 is described in the following: in mode 1, the total data stream from source coding is distributed amongst 1536 carriers. Each carrier is DQPSK modulated with modulation of all carriers made on the same clock, and the sum signal is transmitted terrestrially.

The phase shift keying of the carriers is performed by a digital pulse not having an S-slope. That is why the spectrum of each carrier follows a sinx\x function and is not limited to a defined narrow band. There are periodic

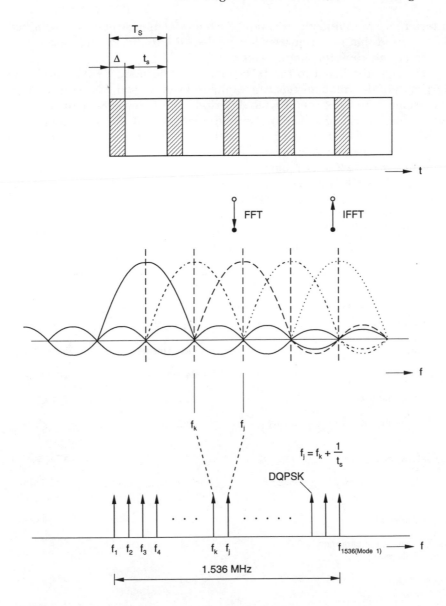

Fig. 2.44 DAB signal in time and frequency domain

nulls in the spectrum which corresponds to the adjacent carriers. This configuration is called an orthogonal carrier arrangement (Fig. 2.44). The dense carrier occupancy leads to an optimal use of the frequency range. The orthogonal arrangement allows a clear separation of the modulation signals in the receiver. The insertion of a guard interval avoids intersymbol interference in multipath reception. Each modulation symbol is transmitted for a period that is one-quarter longer than the useful symbol period that would be sufficient for simple transmission.

The signal radiated by the DAB transmitter in mode 1 consists of 1536 carriers and the frequency spacing of the individual carriers corresponds to the reciprocal of the useful symbol period. The latter is 80% of the total symbol period calculated from the frame length and the total number of symbols. The allocated guard interval is longer than the maximum delay assumed in terrestrial single-frequency networks. The total bandwidth transmitted is 1.536 MHz, approximately 1.3 Hz being required per useful bit. This, however, only corresponds to the spectral efficiency of the RF signals, with the protective bands on both sides of the signal not being taken into account [100].

EUREKA 147 DAB has three defined modes which depend on the network configuration (Table 2.9) [91, 92]:

DAB	Mode I	Mode II	Mode III
Frequency range	< 375 MHz	< 1.5 GHz	< 3 GHz
Application	SFN	local coverage	satellite
T_F frame duration	96 ms	24 ms	24 ms
T_{Null} null character	1 ms	250 μs	250 μs
T_S total symbol period $(= t_s + \Delta)$	1.25 ms	312,5 μs	156.25 μs
t_s useful symbol period	1 ms	250 μs	125 μs
B_F frequency bandwidth $(= N_{OFDM} \cdot 1/t_s)$	1.536 MHz	1.536 MHz	1.536 MHz
Δ guard interval	250 μs	62.5 μs	31.25 μs
N_{OFDM} number of carriers (with QPSK mod.)	1536	384	192
B_s bits/step $(2 \cdot N_{OFDM})$	3072	768	384
F_S step frequency $(= 1/T_S)$	800 Hz	3.2 kHz	6.4 kHz
B_R gross bit rate $(\approx B_s \cdot F_s$ - ref.symbol)	2.4 Mbit/s	2.4 Mbit/s	2.4 Mbit/s

Table 2.9 DAB parameters in modes 1, 2 and 3 [91, 92]

- Mode 1 is specified for the use in single-frequency networks for a carrier frequency up to 375 MHz (broadcast Bands I, II and III).
- Mode 2 is designed for local broadcasting and also includes the local single-frequency network. The RF frequency may be up to 1.5 MHz. This mode is suitable for all TV and sound broadcasting bands (Bands I to V).
- Mode 3 has been developed for terrestrial and satellite configurations and a hybrid satellite/terrestrial application. Investigations have shown that the guard interval is adequate for the multipath profiles in these applications and the mode suitable for RF carrier frequencies up to 3 GHz [95].

2.5.2.4 OFDM generator

The OFDM generator contains the coder and modulator functions. The MUSICAM-coded baseband signals are applied at the input, as shown in our example with six stereo signals coded with 96 Kbit/s per mono signal. Together with the reference symbol a net bit rate of approximately 1.2 Mbit/s is obtained. At the output of an average channel coding of half rate, there is a gross bit rate of 2.4 Mbit/s. The subsequent OFDM modulator (Fig. 2.45) uses IFFT (inverse fast Fourier transform) to produce an I/Q signal which represents the 1536 DQPSK modulated carriers of the OFDM packet. A 1.536 MHz OFDM signal is derived by D/A conversion of the I/Q signal and subsequent I/Q modulation.

In the following, IFFT is to be given a closer look: the input register may be considered as a memory for 2048 equidistant spectral lines, 1536 of which are active. These 1536 carriers are DQPSK modulated. During the symbol period of 1.25 ms (= 1.00 ms useful symbol period + guard interval) the phase relationships are constant. The Fourier transformation is best carried out with the aid of IFFT using signal processors. Within a millisecond the signal processor calculates the real and imaginary components of the time signal samples resulting as a binary form at the output. They are read out during the symbol period and applied to the I/Q modulator. The I/Q modulator consists of two double-sideband AM modulators (DSB-AM) with carriers shifted by 90°. The output signals are taken via an adder to the analogue COFDM signal.

As shown in Fig. 2.46, the I/Q modulator produces in our example a spectral line. The OFDM generator supplies 1536 of these spectral lines which are repeatedly refreshed for the symbol period to yield the OFDM signal with DQPSK modulated carriers.

Notes:
In a digital approach to the I/Q modulator the equivalent of the fourfold sampling rate is used as the carrier frequency. For the cosine component of the carrier the binary sequence is 1, 0, –1, 0... and for the sine component 0, 1, 0, –1... .

The modulation in the I and Q branch and subsequent additions are possible through multiplication by 0, 1 and –1, i.e. it can be illustrated by a

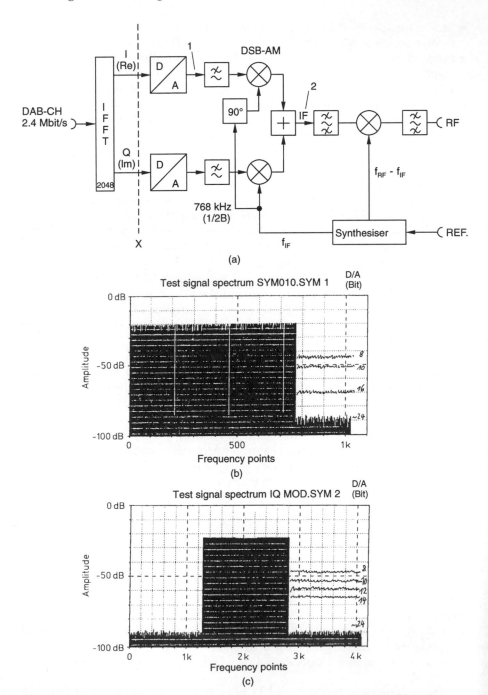

Fig. 2.45 OFDM modulator (x: I/Q interface, for instance with 2 × 8 bit resolution and 2.048 MHz clock)

switchover between the two paths and in the −1 cases by inversion of the output value. To obtain for each multiplier of the carrier signal the corresponding value of the IFFT, an 8 k IFFT and D/A conversion with an 8.192 MHz clock rate have to be performed.

The alternative in the form of an interpolation algorithm with convolution of the time signal via a lowpass filter is far more complex and inaccurate.

Using a SHARP LH9124 chip (Section 4.7.3) [241], the 8 k IFFT can be calculated in four sweeps in 820 μs with the aid of a mixed radix algorithm. Digital I/Q modulation with multiplication of the oversampled time signal takes 205 μs, and the total computing time is within the symbol period of 1.25 ms (mode 1) [245].

The digital solution avoids the disadvantages of the I/Q modulator, where the D/A conversion, alias filtering, mixing and addition of the signals

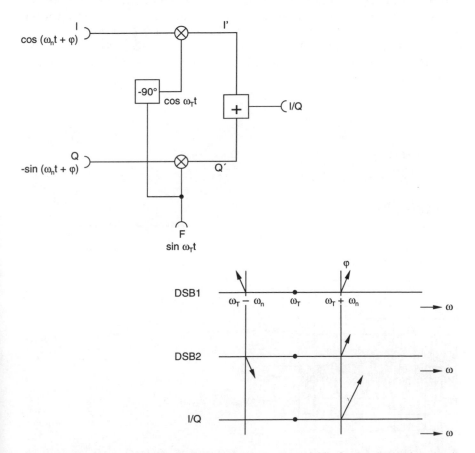

Fig. 2.46 Principle of I/Q modulator (path inversion yields the carrier at $(\omega_T - \omega_n)$, half the bandwidth of the I/Q signal being thus required for the I and the Q signal

from the in-phase and quadrature path, which follow the IFFT, have to be made symmetrically, involving complex alignment and control processes to suppress unwanted signal components.

With digital modulation, the conversion into the IF yields an unwanted sideband which can easily be suppressed by SAW filters.

2.5.2.5 Transmitter amplifier
The OFDM signal has an exact defined frequency, but as a time sum signal has the character of a noise signal (Fig. 2.48).

Due to the DQPSK modulation and fixed phase relationship of the individual carriers, a practical crest factor (ratio of peak to average thermal power in calorimetric measurement) of 8 to 10 dB is obtained. Linear power amplifiers for DAB have to be designed for about ten times rated power. The amplifiers must feature a high degree of linearity since nonlinearities may cause harmonics and sideband signals [90]. The elimination of out-of-band signals by means of bandpass filters is rather complex since the power filters — due to the close spacing of the OFDM packets of 0.2 MHz — must have steep edges and therefore have a relevant insertion loss at the power level. For DAB transmitters a shoulder distance of 40 dB to 50 dB is required (Fig. 2.47).

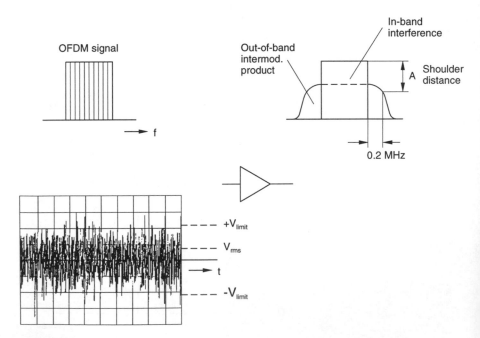

Fig. 2.47 Effect of nonlinear (limiting) amplifier on the OFDM signal

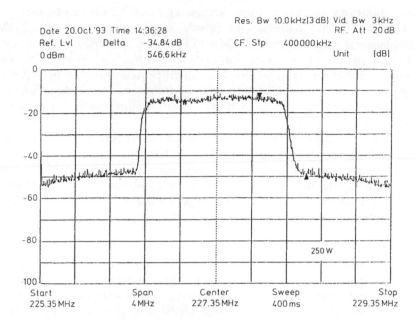

Fig. 2.48 Example of DAB RF signal at the output of a solid-state power amplifier in
class AB mode with precorrection (sync. peak power of TV signal 1.4 kW)

Under these conditions, a typical DAB transmitter power is 1 kW ERP.
With an antenna gain of 6 dB, an antenna input power of 250 W is obtained.
Considering the attenuation of the antenna cable and the steep-edged
channel filter that may be required, a DAB power of 300 W to 500 W must
be provided at the amplifier output (Fig. 2.48). Positive effects such as the
total power amplification in the single-frequency network and field strengths
below the minimum level that are tolerable for a short time (e.g. in 1% of
the time field strengths may be below 50 dBμV) are disregarded. For
specifying DAB power in terms of the mean thermal power of the OFDM
signal it is necessary to state the shoulder distance as shown in Fig. 2.47.

2.5.2.6 Transmitting antenna

Some of the main characteristics of vertical polarisation antenna systems for
TV Band III are described below [101].

These omnidirectional antennas are intended for mounting on mast tops
by stacking several dipoles in parallel. Several dipoles can be arranged in a
selfsupporting GRP (glass fibre reinforced plastics) tube. With four dipoles
a gain of 6 dB is obtained, for a λ/2 dipole with power handling capacity of
1 kW.

Transmitting antennas in anti-icing steel construction with up to four
arrays can produce directional and omnidirectional patterns. With an

arrangement of four bays and four arrays, a gain of 7.7 dB can be met with the omnidirectional pattern and power handling capacity of 32 kW. Combinational antennas for VHF FM and DAB, e.g. channel 12, in anti-icing steel construction are of special interest. In existing VHF antenna systems the conventional dipole arrays for horizontal polarisation can be replaced by dipole arrays with additional vertical radiators for DAB. With this combination, the DAB antenna system does not require any additional space on the mast.

Different radiation patterns can be obtained by feeding the VHF and DAB antennas via separate splitters. The dimensions of the combinational antennas corresponds to those of the previous VHF arrays, only the wind loading will be slightly higher because of the additional radiators. With an arrangement of four bays and four arrays, a gain of 10 dB is achieved for omnidirectional patterns and a power handling capacity of 16 kW.

2.5.2.7 Filters and diplexers
If the DAB signal, for example, in channel 12, has to be supplied together with a TV signal in Band III and fed to a common antenna, a conventional diplexer may be used. The frequency spacing between the DAB and the TV signal must be at least two channels (e.g. channel 10 plus channel 12) [101].

Six-cavity bandpass filters, e.g. for channel 12, with dimensions of 90 cm × 80 cm × 50 cm (W × H × D) are available from Messrs. Spinner [98]. The passband of the filter is 1.5 MHz, with a return loss of greater than 26 dB. At the 2.5 MHz bandwidth points the attenuation is 20 dB, while at the 3.0 MHz points it is 30 dB.

2.5.3 DAB coverage

2.5.3.1 Single frequency network
For the uniform coverage of a country, four OFDM channels are normally required to be arranged in a 7 MHz wide TV channel (Fig. 2.48). Neighbouring broadcasters are each allocated a frequency band corresponding to an OFDM channel to avoid interference between the individual transmitting stations. Within a service area, all transmitters operate at the same frequency, but with the same OFDM channel only being used in nonadjacent service areas (Figs. 2.49 and 2.50). A TV channel is split into four DAB frequency bands of 1.5 MHz each and spaced 200 kHz distance apart.

If in a single-frequency network gaps occur in the coverage of an area, these can be filled by additional transmitters operating at the same frequency. Fig. 2.50 shows that, for instance, the characteristic of a low-power transmitter at the boundary of a service area has a special shape. Topographical conditions may lead to an insufficient coverage, for example, in mountain valleys where gap-filling transmitters solve this problem.

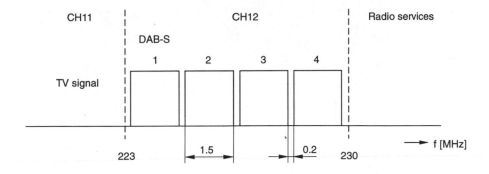

Fig. 2.49 Example of a TV channel occupied by four DAB channels

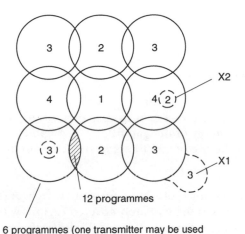

Fig. 2.50 Principle of single-frequency network with four DAB channels for six programmes each (re. X1 and X2; see text)

The gap-filling transmitters need not necessarily be synchronous with the basic transmitters in the single-frequency network. Gap fillers (X1 in Fig. 2.50 and Fig. 2.51) may be configured so that the signal is received at a favourable point, amplified and radiated by a transmitting antenna in the direction of the area to be covered. The isolation between the receiving and transmitting antenna should of course be greater than the transmitter gain to avoid the use of an oscillator. Another effect is the propagation difference between the reradiated signal and the original signal. If the propagation difference does

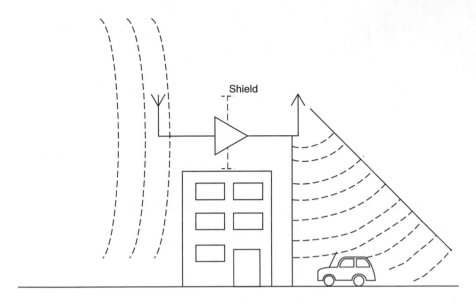

Fig. 2.51 Principle of gap filler

not exceed the guard interval or if in case of a difference greater than the guard interval the energy of the gap filler is negligible, no problems will be encountered with the use of gap fillers in single-frequency networks [100].

2.5.3.2 DAB degradation
In the boundary regions of single-frequency networks or DAB single transmitters, as well as in inadequately covered regions within a service area, reception quality will be degraded. Fig. 2.52 shows the subjective weighting of the reception quality as a function of C/N (carrier/noise). The weighting scale extends from none to little interference (weighting 1) through to strong interference (unacceptable, speech unintelligible at times and frequent muting). Investigations on reception quality were carried out by the IRT using the following parameters:

- error protection profile for convolutional code (UEP: R_{ave}=1/2)
- error concealment for scale factors using DAB-SCF-CRC (cyclic redundancy check)
- carrier frequency = 530 MHz
- MUSICAM 2 x 128 Kbit/s
- transmission channel FADICS (channel simulator from Grundig)
- driving speed = 40 km/h.

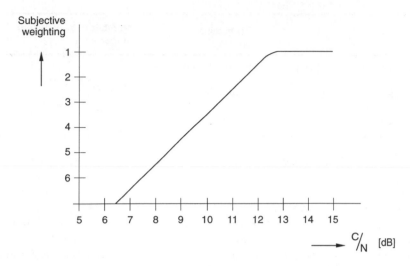

Fig. 2.52 Degradation characteristic of DAB (source IRT, linearised)

2.5.3.3 Regional switchover

In conventional VHF FM networks, a transmitter can be switched from national to local broadcasting. This is of course not possible in a single-frequency network since a temporary switching to subregional broadcasting would cause interference between the transmitters. To overcome this problem, the national programme is continually broadcast in an OFDM channel and provided with additional switching information, in the form of a discrete carrier oscillation on the edge of the OFDM packet, and the receiver is informed that a regional programme is now broadcast on another OFDM channel. With this switching information, the receiver can be switched to local programmes either automatically or manually by the user. The DAB transmitter broadcasts the national signal in an OFDM packet with a specified power in the single-frequency network simultaneously with the regional programme (X2 in Fig. 2.50) with reduced power on another OFDM channel.

2.5.4 DAB transmitter frequencies

When a new broadcasting service such as DAB is introduced, the question of the transmitter frequency is given top priority. Since the frequency demand for individual radiocommunication (mobile radio) is very large and important, it would clearly be advantageous to share the allocated broadcasting spectrum with the new service. In selecting the frequency it has to be considered that the permissible spacing between transmitters decreases with increasing frequency, and the number of transmitters required increases by a power of two (Table 2.10).

Mode	f	Transmitter spacing	Number of transmitters in mode I
1	<375 MHz	approx. 60 km	1
2	<1.5 GHz	approx. 15 km	16
3	<3 GHz	approx. 8 km	64

Table 2.10 Mode-dependent transmitter spacing and number of transmitters

2.5.4.1 Band I (47 to 68 MHz)

In Europe this band is used by the terrestrial television networks (channels 2 to 4), and upon special agreement by fixed radio services and land mobile radio services (military, WARC 1979). Use of Band-I frequencies is expected to be restricted because of transhorizon effects (sporadic E-propagation). These effects occur in particular if high-power transmitters operate in this band covering distances up to 2000 km. Also, Band I is prone to external noise, for example, electrical interference from an underground railway system [86].

2.5.4.2 Band II (87.5 to 108 MHz)

Band II is reserved for VHF FM terrestrial sound broadcasting but it is evident that DAB will be allocated to this band. Since Band II is becoming fully saturated, a start position is required for DAB. Band II is ideal for terrestrial digital sound broadcasting, not only because it has been assigned to terrestrial broadcasting but also because of its position being centred about 100 MHz. The range from 104 to 108 MHz in Band II has already been allocated according to the 1984 VHF FM Frequency Plan of Geneva, and is also densely saturated, mainly by private radio stations.

2.5.4.3 Band III (174 to 230 MHz)

This band is also suitable for terrestrial digital sound broadcasting, although currently it is used exclusively for television, with the exception of channel 12 (223 to 230 MHz), which is mainly used for local TV relay transmitters. Channel 12 is also used by the military services so therefore has only been released for low-powered TV relay or local transmitters. In many parts of Europe (Germany, Italy, Austria, Switzerland, the UK) extensive field trials are planned or now underway using the channel to possibly provide a launching pad for DAB.

2.5.4.4 Band IV/V (470 to 790 MHz)

Over 80% of TV coverage is provided by this UHF band. Channels 62 to 69 are still allocated to the Allied Forces in Germany [86] but there is a good chance that these channels will be released for public broadcasting. In addition to DAB used for regional and local coverage (private programme providers) there are also local commercial TV stations competing for these channels, with terrestrial digital television being a third option for their use.

The UHF frequency range is an acceptable proposition for DAB provided that certain conditions, such as small-cell transmitter network — transmitter distancing of less than 15 km — and a certain speed limit of the mobile receiver, are complied with.

2.5.4.5 L-band - WARC '92 allocation (1452 to 1492 MHz)

At the World Administrative Radio Conference 1992 (WARC '92) the frequency range 1.452 to 1.492 GHz was allocated worldwide to primary user sound broadcasting services (with the exception of the US and Mexico). This was to enable wide-area sound broadcasting coverage via satellite for portable reception and in-car radio reception.

To protect the present services using this frequency range, a rider to the WARC '92 decision stipulates that until the year 2007 the allocation to sound broadcasting has secondary status only. Until 1998 only the frequency range 1467 to 1492 MHz can be used in agreement with neighbouring countries. In 1998 a conference is planned to clarify the situation regarding satellite sound broadcasting.

The WARC '92 decision has brought about changes regarding the DAB frequency. There is no doubt that the original plan to start the DAB service in 1995 cannot be implemented in the L-band. Moreover, extensive investigations of propagation conditions, single-frequency capability and mobile reception criteria have to be completed and many other questions clarified in finding out whether the L-band is really suitable for DAB. The EBU points out that uniform DAB coverage by digital single-frequency networks in the range about 1.5 GHz is technically feasible, but the number of transmitters would have to be increased by a factor of 16 against VHF transmitters (e.g. channel 12). The costs for a terrestrial transmitter network will also be much higher for DAB in the L-band. Another problem is that the DAB receiver for car reception needs two antenna systems, one for VHF and the other for 1.5 GHz.

Utilisation of the L-band for digital sound broadcasting with mobile reception is mainly intended for direct broadcasting satellites (DBS) with terrestrial gap fillers while some organisations are only planning to start with terrestrial DAB in this frequency range [102].

2.5.4.6 S-Band (2300 to 2600 MHz)

To achieve the same coverage as the L-band, the satellite transmitter must have about four times greater power for the S-band. Due to the larger fringe areas, a considerably greater number of gap fillers would be required for S-band coverage. The efficiency of the gap fillers operating at the same frequency would also be reduced by the Doppler effect. Overall the S-band is a very expensive alternative for digital audio broadcasting with very limited technical capabilities.

2.5.5 DAB receiver

Looking at the DAB receiver shows that the DAB functional blocks of the OFDM generator are repeated in the form of a decoder (Fig. 2.53). At the antenna, the DAB receiver sees the OFDM signal with, for example, six programmes. In the boundary regions of a single-frequency network for national programme coverage there may be two or four OFDM signals of different frequencies, but all lying, for instance, within TV channel 12. For programme selection, the DAB tuner must be tunable to one of the OFDM frequencies. In the OFDM decoder, the reciprocal steps of the OFDM generator are processed. DPSK demodulation is followed by MUSICAM decoding and analogue processing for stereo output.

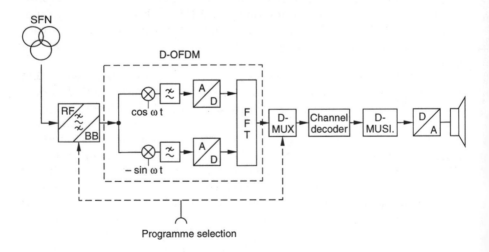

Fig. 2.53 Block diagram of DAB receiver

The requirements placed on a DAB test/measurement or monitoring receiver are more exacting: after decoding, the receiver must make available the DAB data with the associated clock at an output, for BER measurement. Moreover, additional circuits are required for determining frame errors and sync. errors.

ITT is working on a DAB channel decoder on the MASC-3000 concept. The core of a DAB receiver can thus be implemented by means of two chips (plus an external RAM). The channel MASC contains the essential decoding parts. To utilise its capacity, the MUSICAM-MASC provides Viterbi and deinterleaving decoding in addition to MUSICAM decoding.

2.5.6 EUREKA Project EU 147 DAB

The DAB idea was taken up by the IRT (Institut für Rundfunktechnik) early in 1980 when initial proposals were made followed by the first tests in 1985. In 1986, the German Ministry for Research and Technology decided in the presence of representatives from European industry and research institutes to start a European research and development project for digital audio broadcasting. In that year, the EUREKA project EU 147 DAB was launched at the Conference of European Ministers in Stockholm. The first phase covered a period of four years from 1987 until 1991 with an expenditure of about £36 million for approximately 360 man years. In autumn 1991 it was decided to continue the EU 147 project in a second phase from 1992 until 1994 with an expenditure of about £20 million for approximately 170 man years. In the second phase, the main emphasis was given to the completion of the system specification, development of integrated circuits and considerations about new communications services supplementary to the broadcasting services (traffic information and management systems, data transmission to specific groups of users etc.) [103].

In 1990, the 'DAB Platform' was founded in Germany to co-ordinate the interests of the participants in a new broadcasting service and develop a strategy for its introduction. Further aims of this committee included the scientific promotion of DAB, publicity and the system's introduction in 1997.

In France an equivalent to the DAB Platform, 'Le Club DAB' was also founded. The French Radio Numérique is also based on the EU 147 project. A channel 12 implementation seems to be rather unlikely, with the L-band solution being given more consideration.

In March 1993 'The UK National DAB Forum' was founded in the UK to serve as the national DAB platform.

ETSI established a 'Joint ETSI/EBU Project Team 20 V' for DAB, their task being to prepare an ETS (European Telecommunications Standard) for digital audio broadcasting, including mobile, portable and stationary reception. The ETSI Project Team is closely co-operating with CENELEC, EUREKA 147 Consortium, ESA (European Space Agency) and EBU, Technical Committee.

2.5.7 DAB field trials

Any specification, particular that for DAB, has to be verified by field trials. Simulation and real operation of a single-frequency network are of vital importance. DAB field trials were and are being carried out by broadcasters in Europe and worldwide with the majority of such tests being based on single DAB transmitters. In the following three typical examples, DAB field trials are described.

2.5.7.1 Field trial 'IRT Munich'

A very practicable solution for the test operation of a single-frequency network was found early in 1992 by IRT (Institut für Rundfunktechnik) and supported by the industry (Fig. 2.54). Three different sites were chosen in Munich for DAB transmission. The DAB programme was distributed using the TV transposer technique: the analogue OFDM signal is generated at the IRT and radiated on TV channels 11 and 29. At the other two sites the pseudo video signal is converted by a modified TV transposer from channel 29 to channel 11 and radiated via linear amplifiers as used by TV transmitters. A

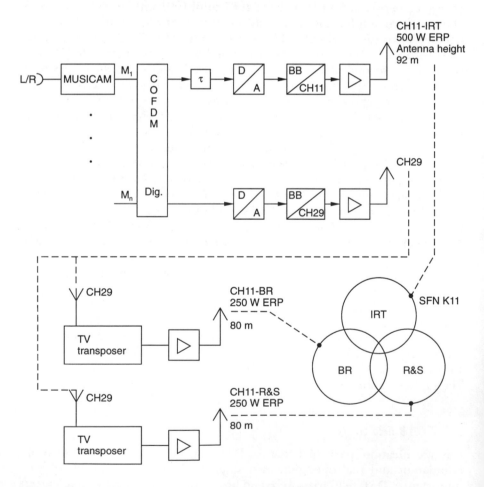

Fig. 2.54 DAB single-frequency test operation in Munich

1 kW amplifier with power tetrode and a 100 W solid-state Band-III amplifier were used and the delay on the path to the DAB transmitter via the transposer of approximately 15 μs was digitally compensated at IRT.

2.5.7.2 Field trial '30/20 GHz Kopernikus'

The satellite is an ideal medium for programme distribution in large field trials and for future operation with regional and national coverage. A point not yet clarified is whether the associated signal delay of about one-quarter of a second will have any adverse effects. A special advantage is the reliability of the transmission medium and the associated standby concept. Since a sufficient number of OFDM generators with coder and modulation functions are not available at short notice and at a reasonable price, it was decided in Germany to use in a large field trial four analogue OFDM signal packets for programme distribution via satellite. The German Bundespost Telekom agreed to make available for this purpose the 30/20 GHz transmission link of the telecommunications satellite KOPERNIKUS from the end of 1992. Fig. 2.55 shows the intended concept: the digital L/R stereo signals from the sound studio are coded with, for instance, 96 Kbit/s per mono signal to reduce the data according to the MUSICAM method. Six data-reduced sound signals M1 to M6 are combined by a DAB multiplexer in a data stream MUX = 1.2 Mbit/s. DAB channel coding with a convolutional code of 1/2 follows so that a data stream of 2.4 Mbit/s is obtained at the output of the channel coder. The subsequent OFDM modulator uses inverse fast Fourier transform to produce an I and Q signal representing the 1536 carriers with DQPSK modulation of the OFDM packet. A 1.5 MHz OFDM signal is derived by D/A conversion of the I and Q signal and a subsequent I/Q modulator. With four such DAB components in total, four OFDM signal packets are produced from a total of 24 data-reduced sound signals. These packets are inserted into a 7 MHz-wide channel and as a combined channel signal taken to the frequency modulator of the 30/20 GHz satellite link. The frequency modulated IF carrier with suitable pre-emphasis is applied via a frequency convertor to the 30 GHz uplink of the satellite.

By this method, four OFDM packets with a total of 24 different programmes can be distributed through analogue transmission to the nationwide single-frequency network transmitters. As shown in Fig. 2.55, each DAB transmitter station has a 20 GHz receiving system, in which the signal received from the satellite is converted into an intermediate frequency, followed by FM demodulation and de-emphasis. A bandpass filter is used to filter out the OFDM packet from a group of four that is to be broadcast from the transmitter station. After delay equalisation and conversion of the baseband signals to the transmitter frequency, for instance, to channel 12 in TV Band III, channel 3 (France) or channel 4 of Bundespost Telekom Berlin in TV Band I, the signals are amplified in the transmitter of the 10 W to 1 kW power class. After passing through a channel filter that may be required, the signals are radiated via the antenna. The bit-synchronous signals radiated by several DAB transmitter stations can then be received with DAB receivers.

This method of transmission is provided by the German Telekom with a minimum of baseband and channel coding equipment being required. An introduced satellite transmission method with frequency modulation and an existing 30/20 GHz transponder of the KOPERNIKUS satellite were used,

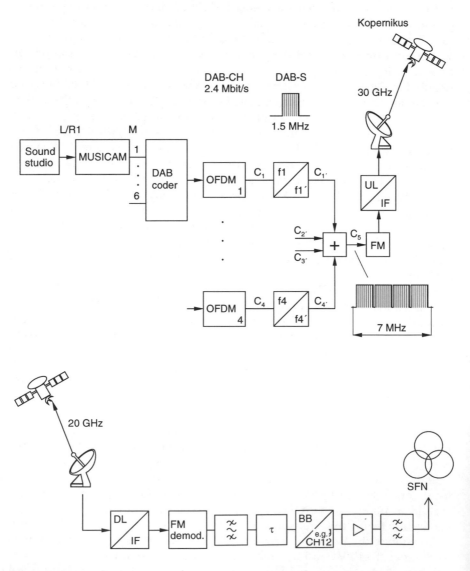

Fig. 2.55 DAB test operation in video mode

although existing equipment components can be employed or adapted for DAB transmitter stations.

Programme distribution via satellite in the so-called video mode, i.e. four analogue OFDM channels with a bandwidth of 7 MHz, and a satellite with a 30 GHz uplink and 20 GHz downlink is a practical method.

However, due to the low efficiency of the analogue OFDM signal distribution and the sensitivity of the 30/20 GHz satellite link with respect to weather-dependent effects, this type of DAB configuration is not considered suitable. Possible solutions are described in Section 4.4.2.

2.5.7.3 Field trial in L-band
The frequency band between 1452 MHz and 1492 MHz allocated to DAB by the WARC 92 may also be shared by terrestrial DAB networks. Encouraging measurements were carried out in Canada in 1992. Early in 1993 CCETT started a field trial in Rennes, the results of which show whether the L-band is technically suitable for terrestrial DAB. In the initial phase, a transmitter with an antenna height of 137 m and a power of 2.6 kW ERP was installed. A coverage radius of about 9 km around the DAB transmitter was expected but the field trial showed the actual coverage with the specified transmitter power at 1.5 GHz [102].

The next stage is the expansion of the coverage area by using extenders, i.e. relays, the coverage diagram of which is adjoining that of the main transmitter. The receiving antenna of the gap filler is therefore aligned to the direction of the main transmitter, while the transmitting antenna is oriented in the opposite direction.

In this experiment, the two DAB modes 2 and 3 are investigated at 1.5 GHz. According to the specification, only mode 3 with a guard interval of 31.25 µs should be used at this frequency. Mode 2 features a greater guard interval (62.5 µs), which might simplify the concept of a DAB network at 1.5 GHz. With the mode 2 guard interval, the transmitter spacing can be expanded from 11 km (mode 3) to 23 km.

France shows great interest in the L-band, since there seems to be no free frequency in Bands I and III (channel 12), and because the L-band is suitable for local networks, serving cities and their environs.

2.5.8 DAB measurements

The subject of DAB measurements, methods and concepts, are dealt with in detail later because the solutions proven for DAB can also be applied to DVB (digital video broadcasting) and digital terrestrial television. These subjects are covered in Sections 3.8 and 4.7.

In the following, the system parameters of DAB mode 1 transmission are assumed for the DAB measurements [105]. Due to the channel model, the following parameters which may influence the coverage are considered:

- the total field strength at the receiver (integral over the bandwidth)
- the field strength and propagation time of the individual subwaves (channel impulse response) and
- the interferences from other sources which may disturb or prevent reception if they are not at a certain spacing from the direct signal.

The bit error rate is a measure for the quality of coverage in which all these effects are combined. Although the BER says nothing about the reason for disturbed reception, it allows a direct conclusion to be drawn about the expected signal quality.

2.5.8.1 Measurement of field strength

Sufficient field strength is a basic requirement for good radio reception. To ensure coverage, the primary aim should therefore be to provide sufficient field strength within the coverage area at any one place and time. Programmes for field strength assessment are an essential planning tool, but results involve a relatively high measure of unpredictability as such programmes are based on rough models. This unpredictability can be compensated for by increasing the transmitter power or the number of transmitter sites. A high transmitter power is not appropriate for DAB because it limits the reusability of the frequency blocks. Moreover, there may be interference in the single-frequency network due to the long-range coverage (guard interval). Neighbouring TV channels or other radio services may be affected too. Therefore, the prediction has to be checked by measurement. Since DAB is suitable both for stationary and mobile reception, two extreme cases should be given a closer look:

- the field strength (as a function of time) at a fixed place
- the position-dependent field strength for mobile reception (as a function of speed).

On the assumption that the total field strength received is a measure of the reception quality, it is measured using the transmission bandwidth of the OFDM signal (1.5 MHz). This assumption is correct if narrowband fading is compensated within the receiving bandwidth by the coding method and interleaving. Missing bits are reconstructed since they only occur individually and not in blocks.

Another requirement is that the delays of the echoes or signals from other transmitter sites do not exceed the guard interval. This condition will certainly be met with a guard interval of 250 µs for VHF. In the UHF range, however, a shorter symbol period, and hence a shorter guard interval, has to be selected, so for high-speed vehicles the interference effects caused by rapid changes of transmission channel within the symbol period, or by the Doppler effect, are minimised. Shorter signal delays than in the VHF range prove to be critical for reception.

2.5.8.2 Receiver for field strength measurement

The OFDM signal has a spectrum which is very similar to that of white noise. It contains a great number of carriers with randomly distributed phases. Therefore, the power within the transmission bandwidth has to be measured. A thermal power meter is   not suitable for reasons of speed. Another possibility is to measure the RMS value of the envelope of the receiver IF signal, with the IF bandwidth of the receiver being equal to the bandwidth of the OFDM signal. This RMS value is a measure of the RF signal power. A short time constant can be selected for the RMS measurement to achieve high measurement rates.

The null symbol has to be observed in the field strength measurement. By synchronising the receiver with the DAB signal it prevents the field strength being measured during the null symbol. Otherwise the measurement would be false since the transmitter does not radiate power during this period.

2.5.8.3 Number of measurements in the field

For a sufficiently high resolution of the site-dependent field strength profile about ten measurements are required per wavelength. This means that at 200 MHz (wavelength 1.5 m) the field strength has to be measured every 15 cm. For a test vehicle driving at 50 km/h this has to be measured approximately every 10 ms to obtain the necessary resolution. The resulting high volume of data is, however, not practical for investigating the coverage area, but for small areas this amount of data is necessary:

• to gain experience for averaging measurements carried out at greater intervals for the whole area to be covered
• for detailed measurements in difficult areas, e.g. in streets with high buildings, and
• to obtain input data for the simulation.

In addition to the reception in vehicles driving at normal speeds, the two most critical cases, i.e. stationary reception and reception at very high speeds, should also be investigated.

With stationary reception, time interleaving does not work if the receiving antenna is within an area of poor field strength. The field strength may vary due to moving reflectors (passing vehicles). These reflectors may contribute to reception so it is of interest to measure the field strength as a function of time.

In the other critical case of very high speeds, field strength is not the only criterion for reception quality because this can be influenced by the Doppler effect.

These two critical cases have to be considered in the definition of the minimum field strength following the measurements performed in the coverage area. Based on the above requirements for a position- or time-

dependent resolution of the data to be obtained, the following requirements
have to be placed on the positioning system in the test vehicle:

- Both a position-dependent trigger for mobile reception and a trigger with
  a fixed clock are required.
- The absolute position determination should be accurate to at least 100 m,
  with 10 m being desirable.
- The resolution over the distance covered should be at least 10 cm.
- With stationary reception, the time resolution should be at least 10 ms.

Due to the low resolution and the shadow effects in urban areas, the GPS
(global positioning system) alone is not suitable for measuring. Suitable
systems would be Travelpilot (from Bosch) for absolute positioning in
conjunction with the Peiseler plate for relative positioning with high
resolution, or the Peiseler plate alone with digitisation of the distance
covered. A combination of GPS and Travelpilot would be a possible solution
both for the city and the country (Section 2.2.4.5).

2.5.8.4 Receiving antenna

The gain or antenna factor of the receiving antenna establishes the
relationship between the input power measured at the receiver and the
receiving field strength. To measure the field strength it is therefore
necessary to use calibrated antennas. A problem, however, is the vehicle on
which the antenna is mounted. The shape of the vehicle as well as secondary
radiators (e.g. windscreen wipers) have a strong influence on the antenna
pattern. Depending on the antenna arrangement, nulls of 20 dB may occur
in the antenna pattern. It is therefore extremely difficult to define a
reference antenna if the vehicle also has to be considered. The recording of
such measurements is nevertheless problematic. On the one hand, the field
strength depends on the direction (nulls in the antenna pattern) and on the
other the vehicle used.

For measuring the field strength, not only are the azimuth patterns of
the antenna mounted on the vehicle relevant, but also the elevation pattern.
This is particularly noticeable in cities where due to reflections the field does
not only arrive from the horizontal direction. Experience has shown that
short rod antennas with their rather homogenous elevation pattern emit
more power in an urban setting than for instance $5/8$-λ rod antennas with
their elevation pattern with sidelobes. In the country it is the opposite, since
in the absence of reflectors the field strength is arriving from the horizontal
direction. For GSM/PCN (groupe spécial mobiles/personal communication
network) mobile radio antennas from different manufacturers are used to
an increasing extent to come as close as possible to real operation. Due to
the strong influence exerted by the vehicle, the definition of the coverage is
crucial. In real operation, 3 dB reserve is added to compensate for these effects.

The rod antenna — mounted in the centre of the car roof — is a suitable
test antenna for vertical polarisation. In future tests other types of antennas

will be tried, such as the disc antenna, for which transmission coverage also has to be ensured.

The consequences are as follows:

- For each vehicle the antenna characteristic and the gain have to be measured in a homogenous field and recorded. Both the elevation and the azimuth pattern should be determined.
- The antenna has to be mounted for as circular a pattern as possible (nulls smaller than n dB, n being determined by the given circumstances).
- For practical reasons, the antenna plus vehicle should not be higher than 2.5 m.
- Measurements on all vehicles must be carried out in identical circumstances and the antenna patterns to be as similar as possible.
- Data comparison becomes possible by measuring the field strength characteristic with different vehicles on the same route.
- Correlation with common types of antennas such as the disc antenna must be ensured.

2.5.8.5 Channel impulse response

The synchronisation symbol which is sent as the first symbol of a frame can be used for measuring the channel impulse response. This symbol is designed so that the transmission function within the channel can be determined. The signal is swept once through the channel. The channel impulse response is obtained by a Fourier transform of the transmission function. Only the magnitude of the individual subwaves is of interest, the phase being irrelevant.

For the field trial, measurement of the channel impulse response is less important than the field strength measurement. In the approved system, the guard interval will be sufficiently long for the reflections to have no significance. Signals from neighbouring transmitters, especially in TV Band I, may cause interference due to the over the horizon effects which occur in this band.

Another application of the channel impulse response measurement is the definition of the propagation profiles for the channel transmitter in receiver measurements and the collection of input data for system simulation [104].

2.5.8.6 Interferences

Another criterion for the operation of DAB is the interference level of external sources. For undisturbed reception, minimum spacings between the DAB signal and the interfering signal have to be defined, with a distinction being made between the different types of interference:

- sinewave interference
- modulated carriers
- pulsed interference.

Sinewave interference and modulated carriers can be measured during the null symbol. Pulsed interference cannot be measured in this way since either the measurement time is too short or interference occurring during the null symbol is not present during the transmission symbols. Pulsed interference can be recognised by a high bit error rate, for which neither the received field strength nor the delays can be the cause.

A test receiver for determining the interference must be able to synchronise with the DAB signal so that the field strength can be measured during the full length of the null symbol. This field strength must be compared with the useful field strength during the remaining symbols. An average value of the power arriving at the receiver may be assumed for the comparison.

2.5.8.7 Bit error rate

The bit error rate measurement includes all the parameters so far described. It provides the most informative data for the determination of the degree of coverage as it completely describes the reception quality. It would be more desirable to make a bit error prediction rate rather than a field strength prediction.

If the BER is too poor, however, no conclusions can be drawn as to the cause, because it comprises several parameters such as field strength, signal delays and interferences, and therefore it is not sufficient to measure the bit error rate alone. If the bit error rate is too poor, all these parameters have to be measured to find out the physical cause. The reproducibility of the BER measurement is strongly dependent on external influences such as passing vehicles and can only be determined by suitable statistical methods.

Another disadvantage is that the measured bit error rate is not independent of the type of receiver used. Depending on the complexity of the implemented error correction method, it may be possible that different values are obtained for the same measurement. Therefore, it is necessary to use a standard receiver against which the other receivers can be referred. This receiver should feature a minimum configuration, since in practical operation coverage also has to be possible for low-cost receivers.

The test rig for measuring the BER contains a combination of transmitter and receiver (Fig. 2.56), the transmitter consisting mainly of a BER generator, an OFDM modulator and a convertor following a power amplifier. The BER generator produces a pseudo noise sequence of length 2^{n-1} which is applied to a DAB transmitter. If a DAB coder is not available, an arbitrary waveform generator may be used instead, which has been programmed with the calculated waveform of the baseband signal (in-phase and quadrature components). The baseband signals are used to drive a transmitter with I/Q modulation capability which converts the baseband signal into RF.

A DAB receiver is required to be switched to a test mode, in which the received data are decoded and the associated clock applied to the output of the BER receiver. For monitoring the measurement and data evaluation it is

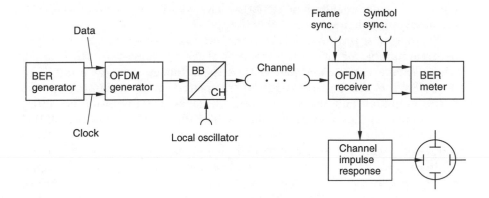

Fig. 2.56 Test setup for measurement of DAB bit error rate

taken into account how often whole frames are rejected as being faulty (frame error), how often the receiver must be resynchronised with the input signal, and how long it takes to synchronise (sync. error).

Another problem in measuring the bit error rate is the time required. Assuming that measurement is carried out in a sound channel with a bit rate of approximately 250 Kbit/s, it takes about 40 s for 100 errors to occur at an error rate of 10^{-5}. In this time, a mobile receiver with a speed of 50 km/h will have covered a distance of 560 m. If the auxiliary data channel is used, the measurement will take even longer. With a measurement time of this magnitude, a mobile measurement is not very practical since an allocation of the measurement specific to a location is no longer possible. The measurement will therefore be carried out at a fixed place, e.g. with minimum field strength, or with a very slowly moving vehicle. However, reception in a car driving at normal speed must be considered.

It would be desirable if conclusions as to the expected bit error rate could be drawn from the level measurement which is fast and made in mobile mode. It is only the level measurement, that can be performed at sufficiently high speed, assuming that no disturbing reflections or over the horizon effects and no interferences are present. In this case the receiving field strength is the decisive coverage criterion in addition to speed.

2.5.8.8 Guard bands
Interference-free DAB reception is only one criterion for good coverage for the listener. Another, but no less important, aspect is that the functions of other services such as television, conventional sound broadcasting and in particular civil and military radio services are not impaired.

Fig. 2.49 shows the problems for channels 11 and 12. Channel 12 is occupied by four DAB packets (1 kW ERP) and channel 11 by a TV transmitter with 100 kW ERP. The second sound carrier of the TV transmitter is close to the channel limit at 223 MHz. The guard band between the DAB transmitter and channel 11 is 200 kHz.

The output spectrum of a DAB transmitter is shown in Fig. 2.48. Due to the great number of different phase-modulated carriers of the DAB signal, a signal similar to noise is obtained whose peak voltage is a large multiple of the RMS value. The linearity requirements placed on the dynamic response of the amplifier and transmitter output stages are therefore far more exacting than is the case with TV signals. The ideal spectrum of the OFDM signal is extended by intermodulation of the individual carriers. At large frequency spacings, these out-of-band emissions can, however, be eliminated by filtering. In the above example the second sound carrier of the TV signal is very close to the DAB channel A. To minimise the spurious emissions, linear TV amplifiers operating far below their maximum power are used in DAB trials but this method is uneconomical for the DAB network of the future.

The level spacings that have to be observed depend on the neighbouring service and have to be defined in trials. A basic distinction can be made between the 'noise to adjacent DAB channels ratio' and the 'noise to other services ratio'. Case A should be less critical, since the modulation method is rather insensitive to interference. In case B, however, very high demands have to placed on the spectral purity of the transmitter signal — depending on the type of the neighbouring service and the level spacing between the DAB channel and its neighbouring service.

To determine the intermodulation products occurring in the transmitter stages the latter can be driven by a noise signal. The conditions are far less favourable than with a DAB signal being used for driving, since the voltage peaks of a noise signal are higher than those of the DAB signal and may therefore falsify the test result.

To determine the guard bands it is advisable to use signal sources which supply signals in line with the final DAB specification. If suitable coders are not available, ARB (arbitrary waveform generators), as for the BER measurement, may be used for signal generation.

After suitable programming, these generators provide the temporal characteristic of I and Q signals in the baseband, simulating DAB signals. These signals are converted into RF by I/Q modulators or appropriate signal generators. According to the programme selected, different signal/noise ratios can be generated and the effects on other radio services thus determined.

2.5.8.9 NMR (noise to mask ratio)

The approach in applying conventional audio baseband measurements to DAB using analogue techniques after the D/A convertor is inappropriate. With psychoacoustics, in particular the resting threshold and signal masking in DAB, analogue measurements and quality criteria such as frequency response, phase response or distortion will prove inaccurate.

Measuring, for instance, a sinewave signal by analogue methods could conclude in judging the transmission quality to be poor, although the music reproduction equates to a live concert.

Therefore, other digital methods of measuring DAB have to be used. One possible approach is by the NMR method ,which emulates the human ear for audio measurements. One of the challenges to DAB audio measurements is the MUSICAM encoder, which can be modified in different ways. ISO Standard 11172-3 contains an exact definition of the filter bank, but for the rest of the encoder section it merely gives a recommendation [93].

The NMR principle uses a method of comparison. The signal to be measured is compared to a reference signal, correlation techniques being used to establish the correct time relationship between the measured signal and the reference signal. The measured signal is always delayed, and the necessary delay of the reference signal is adjusted for maximum correlation. In mobile use, for instance, it is possible to synchronise a stored reference signal with the received signal.

The most important statement is the NMR, expressed in decibels. This is a criterion for whether and to what extent components are contained in the signals that are above the masking threshold (audibility threshold as a function of frequency) and are perceived as a subjective loss in quality. According to the Fraunhofer Society, which has developed the NMR method, the NMR measurement results highly agree with the results of listening tests. The NMR measurement hardware is based on a 32-bit DSP system with three DSP32s (from AT&T).

The NMR method is intended for measurements within the JESSI-DAB Chip project, but are also for use in field trials and coverage measurements in evaluating the subjective transmission quality. For the evaluation of the DAB chips and testing audio coder equipment, an 'objective' automatic testing method will be required in the future.

Similar to the insertion test signal technique in television and the audiodat system for FM sound broadcasting and for TV sound, a monitoring system can be implemented on the basis of NMR. A DAB test signal or reference signal of 0.5 to 1s duration is specified for this purpose and inserted cyclically or in the programme pauses. By sending a start and stop sequence together with the test signal, the DAB receiver can be muted during the test phase. The DAB test signal can be evaluated at any point of the transmission link, using the NMR method, for instance. Using tolerance criteria, a monitoring system incorporating alarm signalling and trend indication by means of statistical functions can be configured.

2.5.9 DR (digital radio) in the US

Despite the numerous and extremely positive field trials held for the DAB system in line with EUREKA 147, the European broadband method has little chance in the US. The probable main reason is that the combining of programmes to form an OFDM packet is against the American practice of broadcasting only one programme from one transmitter station. Moreover, a new frequency allocation for DAB and the granting of the associated licences is considered unrealistic. In the US it is not the technical performance of a digital transmission system that is of prime importance, but its high profitability compared to that of the existing systems. It is even feared that FM and AM sound broadcasting could lose quality on the introduction of DAB. In the US, digital radio is likely to survive only if it supplements the existing sound broadcasting platform rather than replaces it. Uniform coverage in the US does not have the same significance as in Europe, and so the argument on the single-frequency network is less convincing.

This is why in the US so-called in-band systems are taken into consideration, i.e. narrowband systems in which a digital programme is transmitted within approximately the FM bandwidth. The local gaps in the FM band are to be used for the operation of these systems, although the basic disadvantages of narrowband systems regarding the multipath and single-frequency capabilities are well known.

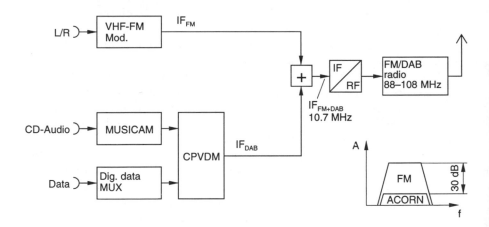

Fig. 2.57 Block diagram of ACORN DAB transmitter
CPVDM = coded polyvector digital modulation

The narrowband system concepts are based on a bandwidth of 200 to 400 kHz. The DR signal is radiated in adjacent channels of the FM stations with a significantly lower power than that of the FM transmitter. This scenario would be suitable for a transition from FM to digital radio and would not require new frequency planning of the current broadcasting spectrum.

A special variant of the in-band systems proposed in the US is ACORN

DAB (IBOC: in-band on-channel). ACORN DAB is based on the MUSICAM audio coding method. The US company Digital Radio specifies a coded polyvector modulation method. The special feature of ACORN DAB is that an FM signal and a digital signal are radiated at the same time by the same transmitting antenna on the same frequency using a co-channel method. The power of the DR signal should be 30 dB below that of the FM signal (Fig. 2.57). This method was presented at the NAB Convention in April 1991 and 1992 as the US's alternative to EUREKA 147. ACORN DAB is a joint development of SRI (Stanford Research Institute), Corporate Computer Systems Inc. and Hammett & Edison Inc. [106]. It remains to be seen whether the narrowband method of utilising the adjacent channel or the same channel as the FM programme will be successful in the US.

Note: Section 2.5 on DAB refers to the basic parameters of the EUREKA 147 system that are necessary to understand its operation. These parameters are described in the references and sources stated (in particular [91] and [92]) and are thus available to the public.

The detailed and comprehensive system definition and specification of the standard (see below) can be obtained by applying for a B-membership with the EUREKA 147 DAB/EBU Consortium.

Digital Audio Broadcasting Eureka Project 147
Primary System Definition prepared by the Joint Eureka 147, DAB W61/ EBU Task Force on System Standardization, Issue 4.0, Dec. 1992

Digital Audio Broadcasting to Mobile, Portable and Fixed Receivers European Telecommunications Standard Preliminary Draft Version 1., Nov. 1992, prepared by EBU/ETSI Project Team on DAB (PT 20-V)

2.6 ADR (ASTRA digital radio)

SES-ASTRA uses 50 — and from 1994 — 64 FM transponders to transmit TV programmes as primary services and sound programs as secondary services. For broadcasting TV sound and sound programmes, ASTRA uses the analogue FM subcarrier method according to WEGENER PANDA (R)-I (companded) with a deviation of ±50 kHz, a bandwidth of 130 kHz and a modulation index $m = 0.15$. The capacity is limited to a maximum of ten sound subcarriers corresponding to five stereo programmes, including TV sound.

The demand for CD quality and further applications, for instance, subscription radio, led in 1993 to the specification of the new digital ADR technique [107].

ADR does not affect the quality of existing television and sound broadcasting services and is compatible with the flexible sound subcarrier concept. For typical 60 cm receiving systems reception is ensured in the 51 dBW service contour.

The system concept is based on these guidelines. It uses audio source coding in line with MPEG-2 MUSICAM with a rate of 192 Kbit/s per stereo signal, including up to 9.6 kbit/s auxiliary data. The QPSK-modulated subcarrier with a modulation index $m = 0.12$ has a nominal bandwidth of 130 kHz. Convolutional FEC 3/4 is used for error protection.

The system tolerance limit with full subcarrier occupancy is approximately 10 dB C/N, corresponding to a BER 1×10^{-5}.

The subcarriers are arranged in the baseband above the video signal (Fig. 2.58) in a 14 division raster between 6.12 MHz and 8.46 MHz and can be occupied by PANDA-I (A_1 to A_6) or MUSICAM-QPSK (D_1 to D_{12}) or in the conventional way as an analogue subcarrier at 6.5 MHz (A_0, mono, not companded, 200 kHz bandwidth, FM, $m = 0.26$).

At 8.595 MHz a special subcarrier (D_0) for network monitoring is used with 14.4 Kbit/s data rate, asynchronous FSK modulation and $m = 0.10$.

The maximum sound subcarrier capacity per transponder is 12 stereo programmes with differential QPSK modulation and two TV sound components in mono or stereo with PANDA-I (alternative 4).

Other possible combinations are:

• analogue subcarrier, 6 × PANDA-I (alternative 1)
• 6 × PANDA-I, 5 × MUSICAM-QPSK (alternative 2)
• 4 × PANDA-I, 9 × MUSICAM-QPSK (alternative 3).

ASTRA refrains from multiplexing MUSICAM-coded stereo signals (e.g. four) prior to the error protection block and QPSK modulation in favour of a clear and transparent broadcast transmission structure.

ADR is particularly aiming at 'digital subscription radio' with consumer-friendly features such as fast programme access and, for example, indication of the title and interpreter/composer as well as of the RDS programme identification (Section 2.2).

ADR is thus competing with DSR (Section 2.3). Moreover, ADR is suitable for audio and RDS signal distribution to terrestrial sound broadcast transmitters (Section 4.4).

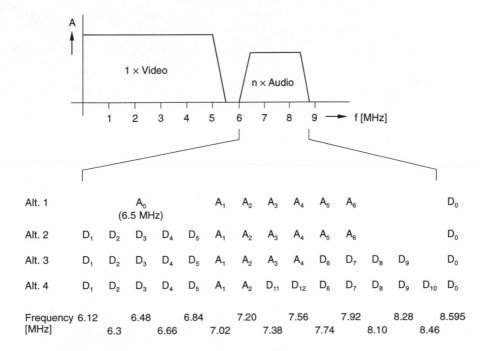

Fig. 2.58 Alternatives for sound subcarrier arrangement on ASTRA satellite transponder

Chapter 3

Advances in television

The international television system standards PAL, SECAM and NTSC have put more development into additional features rather than picture technology (Fig. 3.1). This has evolved during the past ten years with the development being based on the transmitted signal's inherent redundancy in both the time and frequency domains. For example, in the mid-1970s Teletext was introduced in various forms throughout Europe.

The ITSs (insertion test signals) in the field blanking interval ensure a high quality standard of TV broadcast services through automatic measurements and monitoring. Predictions as to the quality can even be made through the insertion test signals and intelligent result processing methods in the form of statistics. Another digital supplementary TV service introduced early in the 1980s was VPS (video programme system) within the TV DL (data line), which ensures videorecorder programmed timer start/stop recordings even in the case of programme rescheduling.

In Germany, TV sound transmission began to improve and be modified in 1980 with the introduction of the DS (dual-sound) carrier method. The digitising of TV sound on programme lines, with the SIS (sound-in-sync) method was used for the first time. NICAM (near instantaneous companding and multiplexing) digitises TV sound which involves a stereo or dual-sound signal being transmitted on a digitally modulated sound carrier in an unused band of the TV channel in addition to the TV picture and analogue sound.

Further development of the TV picture transmission began in the mid-1980s with the Japanese MUSE method being continued in Europe by the compatible MAC method. D2-MAC is a combination of separately transmitted compressed luminance and chrominance signals and a digital sound and data burst. HD-MAC is the compatible version in high-definition quality with this method being further developed to true analogue HDTV. An intermediate step is the analogue PALplus method which is compatible with PAL but has been specified for the new 16:9 format. Experts are, however, rather pessimistic about the future chances of the analogue line of development. Although the 12 Ministers of Posts and Telecommunications of the then European Community member states decided in late 1991 to adopt the D2-MAC standard for satellite television, the MAC project originally developed with the aid of considerable funds from the European Community is generally not seen to have future prospects. The future, no doubt, rests with digital television, which is still in the planning phase, although the US is blazing the trail for digital terrestrial television. In 1993 Europe began to deal with the project of digital television in the terrestrial 8/7 MHz channel (DHDTV terr.).

Digitisation of picture transmission began in the studio with the 4:2:2 component standard specification.

The TV home receiver industry followed with the development of the IC set DIGIT 2000 from ITT Intermetall (DRx: Digital Receiver).

The digital transmission of high-definition pictures via broadband cables (DCATV) or via satellite (DSat) is not a great challenge in technical terms since both the necessary bandwidth is available and the channel response regarding the carrier-to-noise ratio (C/N) is noncritical.

The problem is presented, however, by DVB channel-compatible (digital video broadcasting), which involves terrestrial digital HDTV transmission and uses the TV channels for several programmes in standard quality. Following the revolutionary change to digital techniques, progress will inevitably lead to DIB (digital integrated broadcasting), with full-motion pictures, sound signals and data services being transmitted via terrestrial networks, satellites and cable communication networks with open, transparent interfaces.

Note: In 1993 the Brussels EU Commission stopped the funds for further development of HD-MAC (the European high-definition television). Besides

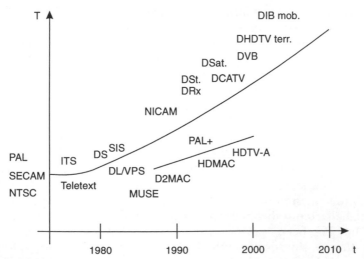

Fig. 3.1 Advances in television

T = technology level

financial reasons, there were technical arguments implying that an outdated technology can no longer be supported. The trend towards digital HDTV is emerging as a world standard. This decision has of course adverse implications on the European D2-MAC format and the Japanese MUSE method which is already in operation. Despite this new specific development, the evolution of analogue HDTV will be included in the following considerations for the sake of a better overview and understanding.

3.1 Insertion of test signals in television

Early in 1980 insertion signal testing in television was introduced in Europe and later in the US, providing the possibility of measuring and monitoring the transmission channel during the on-air programme. This methods allows immediate error diagnostics in the transmission network creating the basis for statistical evaluation allowing the degrading quality or defects to be detected early on.

As insertion signal testing is an analogue method, what is its contribution towards the revolution in television?

The insertion test signals continued the development of programme-accompanying data in television which began in 1975 with the introduction of Teletext. This technique also made use of the free spaces in the transmitted signal. Evolutionary development then came to an end and made way for revolutionary development.

TV measurements and monitoring have ever since been part of TV broadcasting and are gaining more importance with the complexity of the TV transmission network and the ever greater demands placed on the technical quality and operational reliability [108].

TV broadcasting equipment covers studios with TV cameras, video tape recording systems, film and slide scanners and sync. signal systems and PAL coder. Next in the transmission chain are the feed lines used by cable and microwave links to the distribution point and from there to the regional TV broadcast transmitters. Topographical conditions may also require the use of additional relay transmitters or TV transposers before the TV picture can be received by the viewer.

Technical requirements prove to be extremely difficult in joint broadcasting efforts of national and international TV programmes (Eurovision, Intervision etc.) — also via satellite links — since the individual sections of the transmission channel for the TV picture use different modulation modes. With cable feeders, the 21 MHz carrier is amplitude modulated with the video signal (TV-21 technique). In 1985 or so this technique was partly replaced by the digital transmission system DAVOS (digital audio-video optical system) which features a resolution of 9 bits and a sampling frequency corresponding to three times the colour subcarrier frequency (13.3 MHz) and resulting total data rate of 133 Mbit/s.

With a microwave link, a 70 MHz IF carrier is frequency modulated, followed by conversion to the frequency from 2 to 6 GHz. The VHF/UHF signals are radiated under particularly demanding technical conditions using the vestigial sideband method. In view of the large variety of systems switched into the transmission chain and links all using different bandwidths and modulation modes, thorough on-line monitoring and accurate measurement of the RF and video parameters at the distribution points is absolutely necessary.

In traditional TV measurements, the amplitude frequency response and the group delay frequency response were used as the two test parameters. This meant simple tolerance charts and transparent adjustments with clearly visible effects on adjacent frequency ranges. Operation in the frequency range, however, does not decide whether an implemented adjustment is optimal for the transmission of TV signals. These signals cannot be described by the frequency response. That is why in TV techniques the transmitted signal is a mixture of pulses, the spectrum of which has typical maxima. Moreover, nonlinear signal distortion, which cannot be clearly perceived in the frequency domain, suddenly gained importance with the introduction of colour TV.

The time domain has therefore always been used to advantage in TV measurement and monitoring. Standard full-field test signals of similar structure as the typical TV picture elements enable checking and evaluation of linear and nonlinear signal distortion using an oscilloscope, and direct conclusions can be made as to the cause of the TV picture distortion. With 100% modulation of the device under test and a combination of linear and nonlinear transmission errors, interpretation of signal distortion becomes more difficult and with decisions made by qualified personnel. Using full-field test signals means that the device under test has to be taken out of service until the complete transmission chain has been successfully checked. The resulting programme downtimes and interruptions cause time delays and extra costs, particularly when using international links. The heavy programme schedule of broadcasters limits the opportunity of maintaining the systems and links using conventional testing procedures. Therefore, the TV measurement by using full-field test signals has been incorporated into the measurement and monitoring technique using insertion test signals. The insertion test signals contain the essential components of the full-field signals, so all measurements can be made. These test signals are however inserted into the programme-free lines of the field blanking interval and can be transmitted together with the main programme signals. An evaluation at the end of the transmission chain allows diagnosis of the whole link to where the place where the test signals were inserted.

The basic idea of insertion signal testing is to insert at the source an error-free reference signal into the live programme signal. This reference signal passes the entire transmission link through to the home receiver to be measured at any point of the transmission link and conclusions drawn on the programme signal quality. It is no longer necessary to close down the transmitter while under test. By continuous monitoring of the parameters for tolerance limits a monitoring system is configured which is able to deliver an error message within a few seconds. The amount of monitoring compared to measuring the equipment while out of service is considerably enhanced. Another benefit of ITS monitoring is that an intelligent evaluation of statistics can be made using microcomputers. Average-value statistics, event statistics or histograms can be made, allowing a certain degree of trend forecasting to monitor, for instance, the ageing of transmitter tubes or the formation of thermal faults in amplifiers.

3.1.1 CCIR insertion test signals

For implementing insertion signal testing it is of vital importance that certain insertion test signals must be valid in all countries exchanging TV programmes and that individual parameters must be uniformly defined. These conditions have been largely fulfilled through the introduction of the CCIR (EBU) insertion test signals (Figs. 3.2 and 3.7) [110].

The insertion test signals are designated as CCIR 17, CCIR 18, CCIR 330 and CCIR 331, the numbers denoting the line of the first or second field of the programme signal into which the test signal will be inserted. The signals are generated so that the average value of lines 17 and 18 is nearly identical to that of lines 330 and 331 in avoiding 25 Hz flicker on the screen of the home receiver.

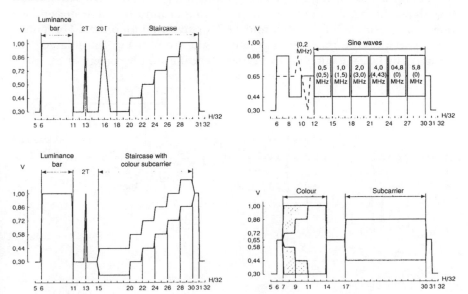

Fig. 3.2 CCIR test signals (17, 18, 330, 331) for TV standard PAL

3.1.1.1 Principle of sample-and-hold circuits

To obtain measured values from insertion signal testing, sample-and-hold circuits are required, at defined times to take samples of the test signals directly or of any signal derived from the test signal. A sample time corresponding to the sampling pulse width of 1 µs is used for almost all parameters obtained. The hold time and the period of the test signal is 40 ms (25 Hz line frequency for PAL). Moreover, the DC voltage results provided by the test signal analysers should be relatively stable even with a poor signal-to-noise ratio. Investigations have shown that a settling or integration time of 2 to 10 s for the test signal analysers is a good compromise between hold time and elimination of interference. Shorter settling times would mean an enhanced sensitivity of the measured values to superimposed noise or interference (interference from other picture or sound carriers, colour subcarrier, pilot frequencies etc).

Around 1980 when insertion signal testing was developed, an analogue sample-and-hold circuit was designed, with the sample circuit having an integration time of 2 to 10 s (Fig. 3.3) [111]. A practically overshoot-free filter of 2nd order with noise rejection improved by about 4 dB as against RC integration (C_1 = 0, R_2 = 0) was used for integration of the results. This sample circuit uses a further essential filter effect. If, for instance, white noise is superimposed on the signal, i.e. all frequencies of the video band have the same amplitude, the interfering frequencies 1, 2, 3 MHz etc. will appear with full periods during the sampling period of 1 µs. Since, however, the integration time constant T_1 = $R_1 \times C_1$ is much longer than the sampling period, these interfering frequencies will be integrated to zero. The dashed line shown in Fig. 3.3 represents a guard ring around the high impedance points between the FET switch and FET operational amplifier. The guard ring is to prevent the measurement result from being invalidated, for instance, by leakage currents from neighbouring supply voltage lines.

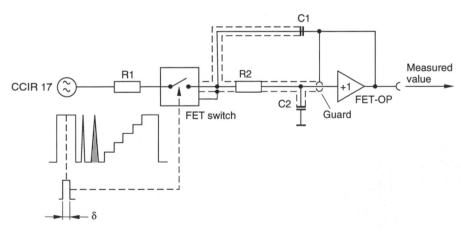

Fig. 3.3 Modified sample-and-hold circuit

3.1.1.2 Test signal analyser
In the simplest case the test signal processor consists of an amplifier, containing possibly bandpass or lowpass filters, rectifier circuits or demodulators. The test signal will normally be sampled at two points and the difference between the two stored values determined. This double sampling principle satisfies the requirement for maximum measurement accuracy with low drift and high noise rejection. This circuit design introduced about ten years ago uses for each test signal parameter a plug-in with two to five sample-and-hold circuits featuring the following data: sample time 1 µs, hold time 40 ms, settling time 10 s. For 25 parameters, about 50 sample-and-hold circuits will be provided in the analyser.

Modern test signal analysers use a completely different technique (Fig. 3.4): there is only one sample circuit and the derived measured value A/D is converted in about 2 µs and before being read into a microprocessor.

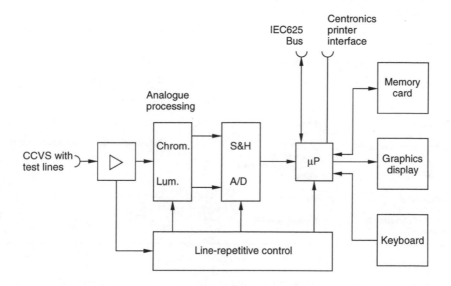

Fig. 3.4 Block diagram of test signal analyser

In the analogue section, the video signal is simultaneously processed by various precision circuits for chrominance and luminance measurements. The output voltages of these circuits with parts of the video signal are digitised by the A/D convertor featuring a resolution of 12 bits and in conjunction with the precision of the analogue circuits ensures high accuracy. A 16-bit microprocessor with arithmetic coprocessor is the core of digital processing. It computes at high speed the results from the sampling values and averages them over the measurement time. Due to the processor technique, stable results can be obtained with integration times of less than 1 s for adjustments with noise-free signals to 10 s with very noisy signals on long transmission links or at the end of transposer chains. The high computing power of the processor allows the continuous computing, updating and monitoring of all test parameters with keyboard, interfaces and memory card also controlled by the processor.

3.1.1.3 Test signal parameters
Table 3.1 and Fig. 3.5 show the relationship between test signal, test signal parameter and the effects of distortion on the TV picture [109] with a test report sample.

Test parameter	Test signal	Line	Effects of distortion on TV picture
Linear distortion			
Luminance bar amplitude	Luminance bar	17/330	Change of contrast
Tilt, streaking, rounding	Luminance bar	17/330	with positive result: streaking with negative result: single-sided plastic effect
2T amplitude	2T pulse	17/330	Loss in contrast for fine details
Colour subcarrier amplitude	20 pulse Multiburst signal 2-level chrominance signal	17 18 331	Colour saturation error, chroma noise
Group delay	20T pulse	17	Colour fringes
Amplitude-frequency response	Multiburst signal	18	
Nonlinear distortion			
Luminance nonlinearity	Luminance staircase	17	Gradation distortion
Differential gain	Superimposed staircase	330	Brightness-dependent colour saturation errors
Differential phase	Superimposed staircase	330	Brightness-dependent hue errors with VTSC (with PAL converted into saturation errors)
Intermodulation	2-level chrominance signal	331	Changes in brightness as a function of colour saturation

Table 3.1 Key test signal parameters and effects of distortion on the picture

```
*** ROHDE & SCHWARZ VIDEO ANALYZER UAF ***
DATE:  11-APR-91      TIME: 14:29:06
MODE: 1       INPUT: A      SYNC: INTERN          MEAS TIME: 1 s
```

PARAMETER	VALUE		LIMIT SET1	ERR SET2	STA-TUS	PARAMETER	DEFINITIONS
BAR AMPLITUDE	-17.1	%	LL	LL		DEF=NOM	....
SYNC AMPLITUDE	18.0	%	UL			DEF=SIG	....
TILT	9.3	%					....
BASELINE DISTOR	1.8	%					....
2T AMPLITUDE	5.0	%					....
2T K FACTOR	0.2	%					....
LUMINANCE NLIN	12.5	%	UL				....
RES PIC CARRIER	13.5	%	UL	UL		DEF=RPC	....
C/L GAIN	-0.3	%				CCIR17	....
BURST AMPLITUDE	15.3	%	UL	UL		DEF=SIG	....
C/L DELAY	2	ns					....
C/L INTERMOD	2.0	%				CCIR17	....
DIFF GAIN (NEG)	-18.4	%	LL	LL		POS:NEG	STEPS=5 ....
DIFF GAIN (POS)	19.8	%	UL			POS:NEG	STEPS=5 ....
DIFF PHASE (NEG)	-4.6	dg				POS:NEG	STEPS=5 ....
DIFF PHASE (POS)	3.9	dg				POS:NEG	STEPS=5 ....
C NL GAIN (NEG)	-4.3	%				POS:NEG	....
C NL GAIN (POS)	0.0	%				POS:NEG	....
C NL PHASE (NEG)	-0.3	dg				POS:NEG	....
C NL PHASE (POS)	0.8	dg				POS:NEG	....
MULTIBURST 1	3.3	%					....
MULTIBURST 2	3.0	%					....
MULTIBURST 3	4.1	%					....
MULTIBURST 4	2.3	%					....
MULTIBURST 5	2.2	%					....
MULTIBURST 6	1.2	%					....
SIGNAL TO NOISE	72.4	dB				DEF=SIG	WGHT=ON FSC=OFF ....
CH/SND INTERMOD	64.9	dB				DEF=SIG	RFCOR=OFF ....
HUM	15.5	dB	LL	LL		DEF=SIG	....
EXTERN DC	-0.002	V					....

Fig. 3.5 Test report of insertion test signal analysis

3.1.2 Test signal monitoring

3.1.2.1 Transmitter station monitoring

A good example of the use of the test signal technique is given by the German Bundespost Telekom while monitoring the transmitter stations of their terrestrial network, and by 1982, all 90 transmitter stations were equipped with test signal monitoring systems. They operate in a decentralised mode, i.e. a separate monitoring system is assigned to each transmitter site. The TV monitoring system comprises a video testpoint selector, a test signal analyser and a data processor for processing the results (Fig. 3.6). At each transmitter station, the video signals with insertion test signals of the second and third TV programme at the transmitter input, and the demodulated RF transmitter output signals of the transmitters A and B, are applied as test signals to the monitoring system. The programme signals are cyclically switched to the test signal analyser. The data processor handles the digital values and status information, and records them in tables as a function of the testpoint. If the status changes or tolerances are exceeded, a corresponding message will be sent via serial data interfaces to the central broadcasting station. Data are transmitted via modems and lines of the PTT

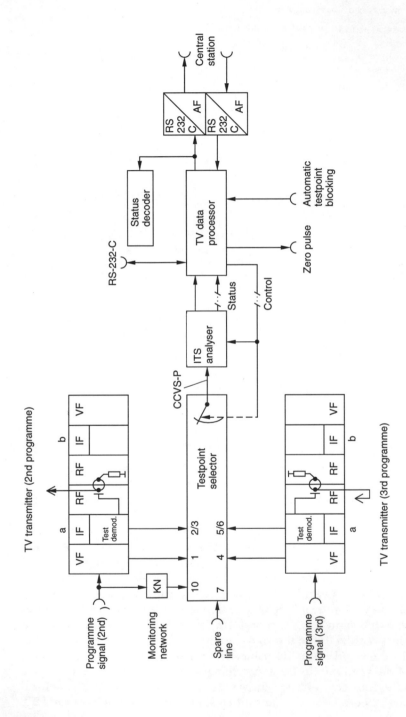

Fig. 3.6 Test signal monitoring of TV transmitter station

network. The status information of the transmitter station is indicated in the central station: changes are indicated by warning displays and have to be acknowledged. With the command structure of the monitoring system, status reports, test reports, average value and event statistics can be accessed. The compressed information about a transmitter station can be obtained using a higher level data structure covering the whole transmitter network. Critical errors, on a specific line network, for instance, can be detected from the general data received.

In parallel with the ITS evaluation, the monitoring system of German Telekom also monitors the TV sound, using a completely different principle. In the sound signal there is neither a gap in the frequency nor in the time domain; the sound/programme signal is used as the test signal. At the programme source, the level of sound signal is measured separately for the left/right information at time intervals of 0.25 s and with 8-bit accuracy, i.e. 64 dB dynamic range and 0.25 dB per bit. The data thus derived are converted into serial form and transmitted at a low data rate of 25 bit/s in line 329, word 5, bit 3 of the TV data line (Section 3.2.1) to the transmitter stations where they are used as reference data. Corresponding to the video programme, the sound signal is analysed similarly to the source signal. By comparing the levels at the destination and at the source separately for the left/right signal, conclusions can be drawn as to the error sources along the transmission link, for example, programme interruptions in the left or right channel, level errors, programme mix-ups or interfering sources along the link.

3.1.2.2 Further applications
Insertion signal testing can also be used in various other applications:

- detection of interference in broadband communication headends or on satellite links
- detection of random errors (e.g. due to weather conditions) as to time and place
- monitoring reception quality at relay receiving stations
- measurements on distribution networks and video routing switchers with different TV standards
- comparative measurements of two video signals for localising losses in quality
- monitoring of transmitter and receiver equipment for servicing intervals
- monitoring of unattended TV transposer stations.

3.2 Video programme system in TV data line

In Germany since the end of the 1970s, and even earlier in the UK, both insertion test signals and teletext signals containing additional information for the viewer have been transmitted together with the TV signals. The need for additional data transmission capacity in the TV signal for TV monitoring and control has resulted in the introduction of the data line technique. By defining the contents of the individual data words data can be transmitted with operational and control data for the video programme system to be processed in special ways.

3.2.1 Specification of data line

According to EBU, the definition of the TV data line transmits information in the form of ASCII-coded words consecutively in each field in addition to a character identification word [114]. In contrast, the data line as specified in Germany offers two separate channels for bit-transparent or coded transmission of data [112, 113].

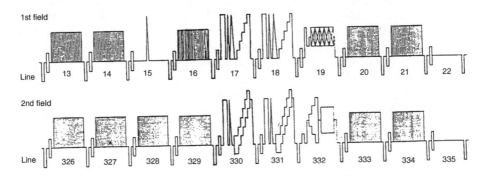

Fig. 3.7 Occupancy of field blanking interval of CCVS (example)

This data line is used mainly for identifying the associated CCVS (composite colour video signal). A distinction is made between the source data line, identifying the actual signal origin to be transmitted unchanged in the first field, and the sectional data line in the second field identifying the specific line section (Figs. 3.7 and 3.8). Moreover, the data line can be used for transmitting sound-related data, for example, implemented as mode identification in the TV dual-sound system, and sound parameters measured with the Audiodat system (Section 3.1.2.1). The transmission of signal-related data (e.g. first, second or third programme, test pattern/programme, news/sport) is also possible. Further applications, such as the plain text transmission in ASCII, transmission of addressed messages and commands, for example, remote control of transmitters and pages of viewer information can be implemented. After processing in the viewer's video-recorder, the additional information decoded by the video

Run-in 10 10 10 10 10 10 10 10 1 bit=2 biphase elements	Start code 10 00 10 10 10 01 10 01	Coded source identification with subdefinition	Plain-text source identification (ASCII sequence)	Sound data
Word 1	LSB Word 2 MSB	Word 3	Word 4	Word 5

Signal content identification programme-related	ASCII plain text network-related	Routing address(es)		Messages/ commands
Word 6	Word 7	Word 8	Word 9	Word 10

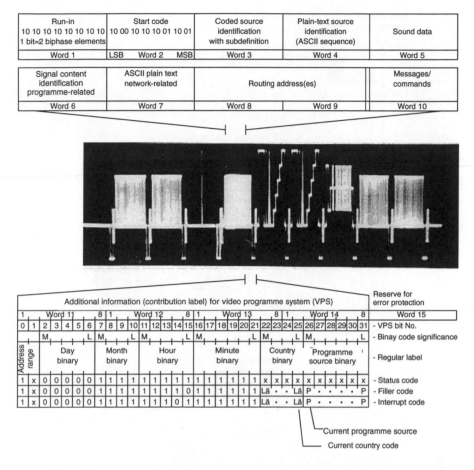

Additional information (contribution label) for video programme system (VPS) — Reserve for error protection

	Word 11	Word 12	Word 13	Word 14	Word 15	
0 1 2 3 4 5 6 7	8 9 10 11 12 13 14 15	16 17 18 19 20 21	22 23 24 25 26 27 28 29 30 31		8	- VPS bit No.
M ─────── L	M ─────── L	M ─────── L	M ─────── L	M ─────── L	M ─────── L	- Binay code significance
Address range	Day binary	Month binary	Hour binary	Minute binary	Country binary	Programme source binary	- Regular label
1 x	0 0 0 0 0 0 1 1 1 1	1 1 1 1 1 1 1 1	1 1 1 1 1 1	x x x x x x x x x x	- Status code
1 x	0 0 0 0 0 0 1 1 1 1	1 1 1 1 0 1 1 1	1 1 1 1 1 1	Lä · · · Lä P · · · · · P	- Filler code
1 x	0 0 0 0 0 0 1 1 1 1	1 1 1 1 0 1 1 1	1 1 1 1 1 1	Lä · · · Lä P · · · · · P	- Interrupt code

Current programme source
Current country code

Fig. 3.8 TV data line with VPS

programme system initiates automatic, timer-controlled recording of the desired programme.

In the data line, 15 byte digital information is transferred as a serial datastream with 2.4 Mbit/s in one line as a result of 15 × 8 bits during the active period of the TV line of about 50 µs. The bit period is 400 ns, with biphase coding being used for the data, that is, the information is coded in a 1-0 or 0-1 bit change. The biphase code is more reliable than the NRZ code, since the receiver can regenerate the clock from the transmitted signal. Supported by the run-in symbol for clock oscillator settling in the receiver and by the start code identifying the data line, the data line is thus independent of the CCVS clock. Despite this fact, within high-quality data line decoders the time position in the field blanking interval is windowed by a blanking pulse. (Practical operation has shown that the run-in symbol and

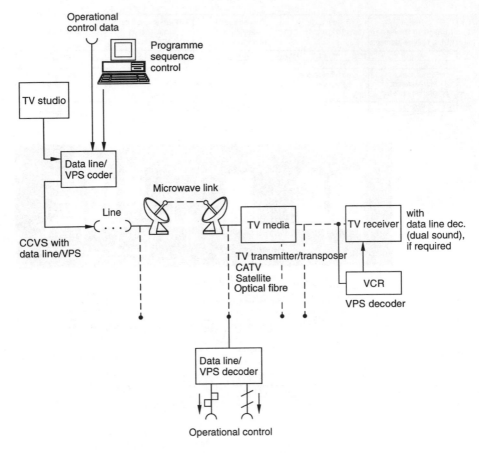

Fig. 3.9 Transmission link for TV data line with VPS

the start code may be simulated by the programme signal and erroneous decoding thus triggered.)

3.2.2 Data transmission link

Fig. 3.9 shows the configuration of a data transmission link using a TV data line coder and decoder. At the studio output the data line is inserted into the CCVS either by a test signal inserter or directly by the data line coder. At the end of each line section a decoder is connected to decode the monitoring and control data [115]. The data line can be regenerated with the aid of a decoder/coder data link at the transmitter station. The useful data are fed to the peripherals via a parallel interface and a serial output channel (RS-232-C). The internally set nominal identification is used for a comparison with the transmitted source or section identification and, in the case of nonconformity, an error signal is set at the message port of the decoder. An additional measurement function of the decoder permits the bit error rate in the ongoing data transmission to be determined and ensures the monitoring of the quality of the transmitted data line signal.

The data line coder can also be used as a VPS generator [116] with the VPS label edited and checked on the terminal. The label is instructed to be inserted into the data line and transmitted at the specified time. If the terminal is replaced by a process controller using appropriate software, it is possible to edit several VPS labels in the external controller in accordance with the TV programme sequence, to enter them into the data line coder consecutively and to activate them by the controller at the corresponding times.

3.2.3 Video program system

VPS is a consumer-relevant service within the TV data line information system. VPS was a logical development after the VCR (video cassette recorder) had become accepted as a basic piece of equipment in practically every home, but timer-controlled programming proved difficult to master. It is an almost everyday occurrence that programme timing differs from the published listings, causing the recording of the programmes to be wrong or incomplete. It became necessary to provide the programme with a label. Only if the data programmed in the recorder (date, time and programme source) coincides with actual data received (labels) will the video-recorder start recording the programme selected and stop when a new or other label is transmitted. Delays in the programme timings become irrelevant since the label transmitted will always contain the correct data including time, date etc.

The recorder must have a VPS decoder for decoding the VPS signals. This decoder compares the data programmed in the recorder with the data of the labels received, and if these are identical it selects the programme to be recorded for as long as the data agrees [117].

In addition to the label for programme-synchronous recording a further identification becomes necessary through the large number of programmes offered by cable, satellite and occasionally TV channels received via community antenna systems which have to be converted to other channels. Additional information has to be transmitted to the TV set providing clear identification of the programme received and on-screen display information.

Words 11 to 14, each comprising 8 bits, in data line 16 are used for VPS (Fig. 3.8). The significance of the 32 bits starts with zero for the first bit of word 11 and ends with 31 for the last bit of word 14. The address range with its two bits can specify four different values. Only the first and second address range are provided for VPS .

Apart from the regular label, the information about day, month, hour and minute is used in the VPS data line for the identification of special status codes.

Since VPS may also be received internationally, information on the nationality of the programme provider can be given in addition to the programme source. The six bits for programme source identification in VPS format allow 63 sources to be identified.

Although not associated to the actual VPS data service, the data contents of word 5 have to be mentioned since the sound status identification is transmitted in this word. Bits 1 and 2 are used to identify dual-channel, mono and stereo status.

Since the introduction of TV dual sound (Section 3.3.1), the information on the sound status is transmitted to the receiver via certain coded frequencies. The receiver industry may also use bits 1 and 2 of word 5 in the data line for this information. Bits 3 and 4 may also become of interest for receiver and recorder configurations. It has been agreed to use these bits for categorising programmes for audience suitability. There are no specific plans yet as to the use of bits 5 to 8 of word 5.

In addition to programme-synchronous recording, VPS should also simplify and facilitate programming of the video-recorder. The programming convenience largely depends on the equipment configuration installed by the manufacturer. Basically, there are three different methods of programming the recorder:

* entering a number sequence corresponding to the information printed in the programme listings
* scanning a bar code printed in the programme listings by a light pen. This method is only available if the publishers of programme listings print the bar code
* the VPT method (video-recorder programming by teletext) automatically takes the necessary information on the programme to be recorded from the teletext index pages on which the desired programme is listed.

3.2.4 VPT system

The programme index teletext pages of the broadcasters provide nearly all the requirements for programming of and data transfer to the video-recorder. The current VPS data are highlighted on the screen in magenta in addition to the scheduled broadcasting times. Moreover, control and monitoring characters some of which are visible have to be denoted in the index pages, with a host of hidden characters which are exclusively used for the data transfer to the VPS computer. According to specification, the relevant VPS data include the programme source, the country of origin, the time and the date of broadcasting. These data are terminated by a guard word for detecting errors in the data transfer by the checksum. Operation is simplified as the channel, programme, and start and stop times are not manually entered by the viewer. All the viewer has to do is to mark the desired programme on the teletext index page, which automatically programmes the video-recorder by decoding the information contained in the teletext pages [118, 119].

Besides the simple and error-free programming, the VPT method has further advantages for the user:

• storage of programme titles
• on-screen display of all stored programmes for recording.

The PDC (programme delivery control) system has been specified at the EBU and ETSI level, supported by the JTC (Joint Technical Committee) co-ordinating the work of these two bodies issued early in 1991 as an ETS (European Telecom Standard). The PDC specification includes and enhances the VPS and VPT specifications. Two different methods can be employed for video-recorder programming by teletext: programming via the TV data line or via the MAC packet signal. This means that PDC also includes the German data line standard regarding the start code and the word allocation, although not in detail.

3.3 TV sound broadcasting methods

3.3.1 Stereo and dual sound in television

Since 1981 the TV medium has been bringing a further innovation to most parts of Europe: stereo sound and dual-sound transmission. The combination of the colour picture and stereo sound presents the viewer with the kind of experience only felt in a concert hall, theatre or cinema. Alternatively to stereo sound, this innovation included dual-sound broadcasts which opened up new possibilities to television: foreign films, for instance, could be broadcast in the native language on one channel and dubbed into the local language on the second, with the TV 'listener' able to choose the language.

After the introduction of colour TV in 1967, further development in TV broadcasting mainly related to additional information services such as teletext, data line with VPS and — with certain restrictions — insertion signal testing as well as TV sound by digital transmission techniques, with one exception: dual sound using the IRT method. In this case a second FM modulated sound carrier is used in the gap between the previous 5.5 MHz FM sound carrier and the channel end. The compatible stereo and dual-sound method became widely accepted in Central Europe since the transmission link including the TV transmitter could be adapted to the analogue dual-sound technique with little extra cost. The dual-sound system based on two FM carriers (Fig. 3.12) was developed and patented by IRT (Institute für Rundfunktechnik) in Munich and is used in Germany,

Switzerland, Austria and other countries outside Europe. The system is technically described in Section 3.3.2.

Regarding TV sound, all the other new methods use digital sound transmission techniques: SIS (sound-in-sync.), NICAM and all MAC methods.

3.3.1.1 Original alternative transmission methods

For implementing dual sound a second sound channel is required between the TV studio and the home receiver and the chosen transmission method must be compatible with the existing standard. This means that receivers complying with the previous standard must still be able to receive mono sound signals without interference from the second sound channel. Three different transmission methods were originally considered [127, 129]:

- Using a pulse code modulation method, two sound channels are integrated in to a time multiplex into the sync. pulse and blanking interval of the video signal (Fig. 3.10). This TV-PCM2 method [123] has the essential advantage of not needing separate sound transmission lines and sound distribution, and therefore a better frequency economy can be achieved. It does not, however, satisfy the requirement for compatibility with existing home receivers.
- With the multiplex method (Fig. 3.11), the second sound signal is modulated onto an auxiliary carrier, similar to the stereo multiplex signal in VHF sound broadcasting. In October 1978 in Japan, the FM/FM transmission method was introduced. With this method, an auxiliary carrier of twice the line frequency is frequency modulated with the second sound signal. The combinational signal of the first sound and the frequency modulated second sound is frequency modulated onto the RF sound carrier. In view of the double frequency modulation of the second sound signal, this method is called the FM/FM multiplex method. A comparison with the dual-sound carrier method described in the following [124] showed that the FM/FM method has disadvantages with respect to almost all parameters of the dual-sound technique, e.g. signal-to-noise ratio, channel crosstalk or distortion.
- With the third method, the dual-sound carrier method (Fig. 3.12) [122, 125], a second frequency modulated sound carrier is inserted 242 kHz above the first sound carrier, being 5.5 MHz above the vision carrier for standards B and G. The two sound channels share almost the same technical data and quality characteristics, with the only difference being that the second sound carrier is radiated with reduced power to avoid adjacent channel interference.

In Germany, the decision was made in favour of the dual-carrier method at the 20th conference of the FuBK (television committee of the German authority for television transmission) in February 1980 with the relevant standard being adopted. The decision supported this method because of the

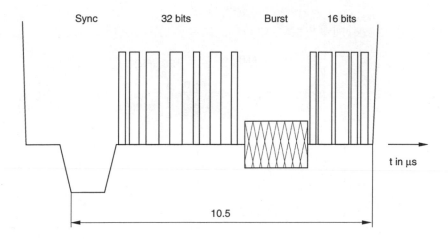

Fig. 3.10 TV-PCM2 transmission in the line blanking interval

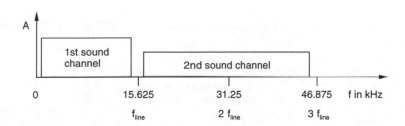

Fig. 3.11 FM/FM multiplex method

high transmission quality in both sound channels, the compatibility with the existing home receivers operating on the previous standard, the resistance to interference from the offset mode of TV transmitters [126] and because of the favourable implementation costs.

The problems encountered with the dual-sound carrier method are the high crosstalk attenuation that is required when two independent sound signals are transmitted together and the high demands placed on equal phase and amplitude in the two channels when stereo sound signals are broadcast. In addition, the more densely occupied TV channel may affect the power levels and intermodulation products of the transmitter.

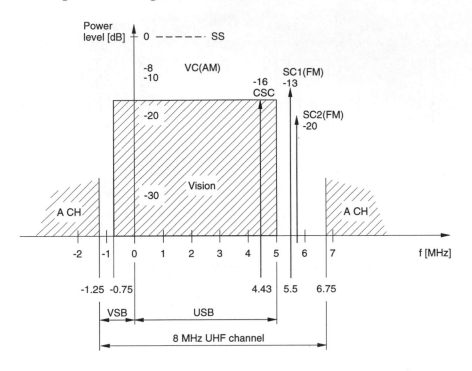

Fig. 3.12 Sound carriers in UHF channel (Standard G, VHF channel bandwidth: 7 MHz)

ACH = adjacent channel   CSC = colour subcarrier
SP = sync. peak   SC = sound carrier
VC = vision carrier   VSB = vestigial sideband
                                USB = upper sideband

3.3.1.2 Dual-sound transmission

Dual-sound transmission begins at the output of the TV studio (Fig. 3.13), providing a second sound signal and control outputs to define the operating mode: mono, stereo or dual sound.

The two sound signals are transmitted to the TV transmitter in a PCM data channel with a capacity of 2 Mbit/s and sufficient to carry five sound channels. The data channel is inserted into the 10 MHz baseband of the microwave link. The current operating mode is transmitted together with the video signal and contained in coded form in TV data line 16 (Section 3.2.1). Line 16 and line 329 in the second field are normally used for transferring internal information such as measured data, remote-control data or information referring to the signal contents.

At the TV transmitter the sound signals received via the microwave link are applied to the TV dual-sound coder providing several functions:

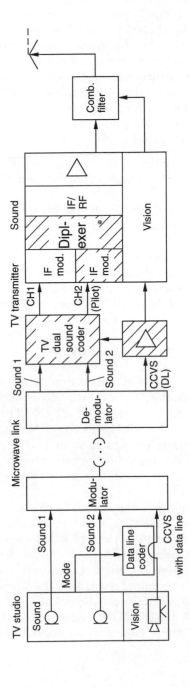

Fig. 3.13 Dual-sound transmission from studio to TV transmitter (crosshatched: additional equipment required for dual sound)

- 15 kHz lowpass filter for band limiting of the sound signals
- matrixing in stereo mode, producing the signal $M = (L + R)/2$ so that a compatible sum signal can be offered to the existing TV receivers
- 50 µs pre-emphasis. In the mono mode, pre-emphasing the high frequencies was previously carried out in the sound modulator. Now it has to be done prior to matrixing, since in the stereo mode asymmetry in the pre-emphasis after matrixing would cause crosstalk
- decoding of the dual-sound mode from the data line and re-encoding by modulating the 54.7 kHz pilot carrier with identification frequencies assigned to the dual-sound mode.

The modulated pilot frequency is added to the AF signal in channel 2. The pilot frequency and the identification frequency are linked to the line frequency.

The dual-sound signal reaches the stereo home receiver, possibly via TV transposers and community antenna or CATV systems.

3.3.1.3 Dual-sound characteristics

The characteristics of the dual-sound standard are summarised in Table 3.2 [120, 121]. At the studio output and in the microwave distribution network the AF lines designated 'Sound 1' and 'Sound 2' are used:

in mono	for the mono signal M1,
in stereo	for the left- and right-hand information L and R, and
in dual sound	for two separate mono signals M1 and M2.

The dual-sound mode is coded in data line 16, word 5, bits 1 and 2.

In the stereo mode the TV dual-sound coder at the TV transmitter input produces the compatible sum signal $M = (L+R)/2$ on the AF line channel 1. while channel 2 carries the right-hand signal. Matrixing the multiplex signal with the sum and difference signals as in VHF sound broadcasting is not adopted here for TV sound. Correlating noise signals as they occur in television would otherwise be additive at one sound output and subtractive at the other.

The dual-sound mode is transmitted in channel 2 using a 54.7 kHz pilot signal remaining unmodulated in mono mode and is amplitude modulated with the identification frequency of 117 kHz in stereo mode and 274 Hz in dual-sound mode. Standard specifications dictate a frequency of 5.742 MHz above the vision carrier for the second sound carrier. The frequency deviation is 30 kHz at nominal level, with the unmodulated pilot signal in channel 2 producing an additional deviation of 2.5 kHz. The vision/sound power ratios of 13 dB and 20 dB were selected to prevent overloading of the transmitter sound output stage and to avoid interference from adjacent channels [128].

	Sound studio			
	Mono	Stereo	Dual sound	Identification
Sound 1	M1	L	M1	Data line 16
Sound 2	M2	R	M2	Word 5, bits 1 and 2

	TV dual-sound coder			
CH1	M1	$M = \dfrac{L+R}{2}$	M1	
CH2 Identification frequency	M1 0 Hz	R 117 Hz $(=1/133\,f_{line})$	M2 274 Hz $(=1/57\,f_{line})$	Pilot carrier 54.7 kHz $(=3.5\,f_{line})$ 50% AM with identification frequency

	Sound transmitter		
	Sound carrier	Frequency deviation	P_{vision}/P_{sound}
CH1	f_{vision} + 5.5 MHz $(=353 \times f_{line})$	30 kHz	13 dB
CH2	f_{vision} + 5.742 MHz $(367.5 \times f_{line})$	30 kHz (+2.5 kHz in mono)	20 dB

Table 3.2 Characteristic data of dual-sound standard

3.3.1.4 Dual-sound TV receiver

Fig. 3.14 shows three alternatives for retrieving the two channel signals CH1 and CH2.

A cost-effective solution is to utilise a second frequency discriminator for the intercarrier frequency 5.742 MHz. This solution, however, does not satisfy the enhanced requirements placed on the sound channels.

The so-called quasi-parallel-sound demodulator provides a technically good cost-effective solution. The intercarrier frequencies for the IF vision carrier and the two IF sound carriers are derived via a separate filter to eliminate interference due to vision modulation.

The best technical solution is obtained by employing a more elaborate configuration. Intercarrier interference can be avoided by using parallel-tone reception. With this method, the frequency is not demodulated via the difference frequency between the vision and sound carriers, but, for instance, directly via the sound IF of 33.4 MHz for sound 1 and 33.15 MHz

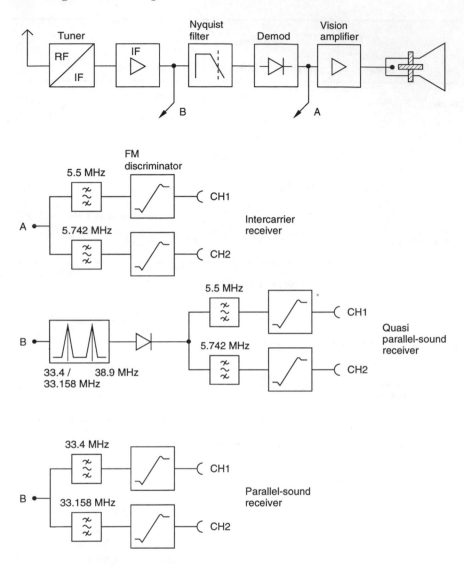

Fig. 3.14 Various sound demodulation methods

for sound 2. The use of low-cost synthesiser oscillators and SAW filters for the sound IF may increase the chances for an application using this method.

Demodulation of the channel signals is followed by dematrixing (Fig. 3.15) which is driven by the identification frequencies from the pilot signal. The AF selector, sound output amplifier and loudspeakers follow.

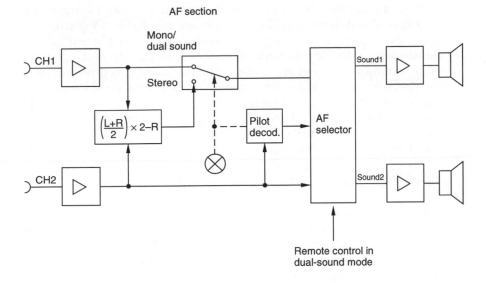

Fig. 3.15 Dual-sound dematrixing in TV receiver

3.3.1.5 Dual-sound parameters

The major quality determining parameters in the dual-sound system are:

- **Intercarrier noise** is produced in both the transmitter and the receiver by spurious phase deviation of the vision carrier, mainly as a result of vision modulation with the 50 Hz harmonics and the 15 kHz components of the video signal.
- **Channel crosstalk** is a dual-sound parameter that takes on an importance when two sound signals are being transmitted, e.g. one in the original language and one with dubbing into the local language. Crosstalk may occur in the AF stages of the transmitter or in the dematrixing circuits of the receiver. Channel crosstalk is also possible in RF stages as a result of synchronous amplitude modulation. In this case, synchronous AM accompanying the frequency modulation of the sound carrier may cause crosstalk on the second sound carrier through intermodulation.
- **Stereo crosstalk** is a parameter which is mainly determined by the type of matrixing, i.e. $(L+R)/2$ in channel 1 and R in channel 2. Differences in the phase and amplitude responses of the two channels cause crosstalk as the balance between matrixing and dematrixing is lost. Asymmetrical frequency deviation also causes stereo crosstalk.
- **Intermodulation of the two sound carriers** with one another and with the vision carrier is produced by nonlinear or overdriven RF amplifiers, e.g. in a sound transmitter output stage, in TV transposers and especially in the amplifiers of community antenna systems. Intermodulation may

produce an interference line located above the vision carrier at the spacing of the sound carrier difference frequency. In the TV picture this causes moiré patterns, i.e. a picture disturbance varying with the AF.

- **Transmission reliability of mode identification:** The current operating mode is transmitted from the studio to the transmitter via the data line in the video signal. A faulty video signal or data line causes automatic switchover to the dual-sound mode in the TV dual-sound coder. In the TV receiver, identification of the operating mode may be impaired by intercarrier interference if vision modulation is performed with three and a half times the line frequency corresponding to the pilot carrier frequency. This interference source can, however, be practically eliminated by providing adequate selectivity in the pilot demodulation.

3.3.2 Sound-in-sync method

SIS (sound-in-sync) is a programme accompanied data transmission method which uses a redundancy in the time domain of the CCVS, namely the 4.7 µs sync. pulse, for digital sound transmission. SIS marks the beginning of digital sound transmission in television and is an important milestone in the analogue/digital development of TV. An essential advantage of SIS is that no extra AF lines are required in addition to the video line, determining that the TV accompanying sound is linked to the picture signal using PCM (pulse code modulation) (Fig. 3.16).

Apart from decommissioning the permanent sound network, direct coupling of the TV accompanying sound to the TV picture signal has also operational advantages. Both signals take the same route and same crosspoints and monitoring points. This would not be possible in conventional networks where signal routing is sometimes very divergent. SIS also brings a considerable improvement of sound quality with lower susceptibility to interference on the transmission links [130].

With the SIS method, the AF signal is sampled with twice the line frequency of 31.25 kHz so that a basic bandwidth of 14 kHz is obtained for sound transmission. Since double the line frequency is used for sampling, there are two samples per line. The samples are binary coded to yield two 10 bit pulse groups. The second pulse group is transmitted in a complement code. By means of time interleaving a group of 20 bits is obtained. The beginning of the pulse group is identified by a marker pulse which defines the start reference for decoding. The information bits following the marker pulse start with the LSBs (least significant bits). This makes for immunity to effects caused by interference in the picture signal (changes in phase/frequency response). The data pulses are bell-shaped similar to the 2T pulse used in insertion signal testing and feature a half-amplitude duration of 182 ns. The total duration of the SIS signal is 3.82 µs. The amplitudes of the pulse peaks correspond to the standard level of the video signal of 1 V_{pp} [131].

For a continuous transmission of the SIS signals in the field blanking

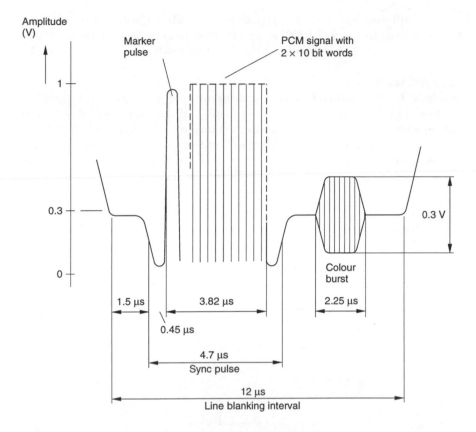

Fig. 3.16 Sound-in-sync signal in line blanking interval

interval of the CCVS, the equalising pulses of the field interrupt pulses are extended from 2.35 µs to 4.5 µs. The extension of these equalising pulses has no effect on the signal transmission via terrestrial transmitters, since the inserted SIS signal is extracted prior to broadcasting.

The SIS system was installed at switching and monitoring points of the EBU network early in the 1980s. Altogether, about 100 of these systems are in use or are made available by EBU to the individual broadcasting organisations. The transition from analogue sound to SIS is effected at the interface and interchange points between the national networks and the international permanent network for vision PN-V (permanent network vision).

The SIS system has been designed for the TV standard in line with CCIR 625 lines/50 Hz and can be used for PAL and SECAM. For standard conversions, transcoding or recording (VTR video tape recording) the SIS signal has to be decoded and the TV sound processed in the conventional way.

The SIS method was developed by the BBC (British Broadcasting Corporation). In the mid-1980s a stereo version was implemented, currently being used in the line networks of the Austrian ORF.

3.3.3 NICAM method

Similarly to SIS, NICAM uses a redundancy in the video transmission channel, but in the frequency domain instead of the time domain. A sound carrier with digital modulation is inserted into the TV channel between the analogue sound carrier and the channel end. In contrast to the dual sound carrier method, the two sound signals form a single digital signal, thus eliminating the problem of compatibility of the second sound channel.

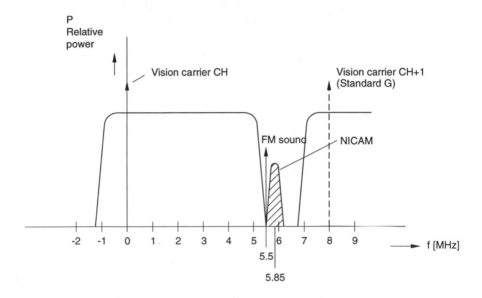

Fig. 3.17 TV channel occupied by NICAM signal (UHF band, Scandinavia, Spain)
Problem: Standard B with 7 MHz channel bandwidth;
Standard I (Great Britain) with USB of 6 MHz at 8 MHz channel bandwidth

NICAM 728 was adopted by Spain and the Scandinavian countries in the mid-1980s, giving Europe a digital multichannel sound transmission in addition to the analogue dual-sound carrier method used by Germany, Switzerland, Austria and the Benelux countries.

The second NICAM sound channel carries the entire dual-channel audio information. If the compatibility requirement is abandoned, the first sound carrier may be used as a third channel for other purposes, for example, the transmission of a third language.

Due to the different channel bandwidths employed in Scandinavia (7 MHz) and in Great Britain (8 MHz), two different NICAM versions have been adopted: the Scandinavian NICAM features a smaller spacing between the sound carriers and a smaller bandwidth of the digital carrier, with the data rate being identical (Fig. 3.17).

3.3.3.1 Coding
The additional NICAM carrier is modulated with two sound signals, i.e. either a stereo channel or two separate mono channels available for dual-language sound transmission. In contrast to the analogue dual-sound carrier method, channel separation in the dual-language mode is particularly noncritical because of NICAM's digital modulation. A channel separation of 80 dB is achieved both in dual-sound and stereo mode.

Fig. 3.18 shows the structure of a NICAM coder: the inputs A and B are analogue audio sources with 15 kHz bandwidth which are first subjected to a pre-emphasis in line with CCITT recommendations. The two analogue/digital convertors use a sampling frequency of 32 kHz and a resolution of 14 bits. The audio data are PCM-coded at the A/D converter outputs using a two's complement code [132].

The data rate of 2×14 bits $\times 32$ kHz cannot be directly transmitted in the available bandwidth; therefore the NICAM coder compresses the data from 14 bits to 10 bits. This technique gives NICAM its name: a contraction of Near Instantaneous Companding. The error protection used relates to that of the MAC method (Section 3.4.2), where digital TV sound was implemented as part of the D2-MAC satellite standard. In the coder, the data are processed in serial form in 64-word packets (frames). One frame corresponds to the audio information of 1 ms per channel. In the stereo mode, the frame contains 32 words from the right-hand channel as well as 32 words from the left-hand channel. In the dual-sound mode, all 64 samples of a frame are assigned to a mono channel, ie one frame contains the input information of one channel of 2 ms. This data information in packets supports the subsequent data compression, for which a scale factor is determined for five coding ranges and coded with 3 bits. The scale factor gives the maximum value of the signal amplitudes of each channel in the time window of 1 ms, corresponding to 32 words. The scale factor is a reference quantity for correlating audio signals and permits the NICAM signals to be represented by 10 bits referred to the scale factor. There is a certain analogy with the DS1 standard for DSR (Section 2.3.2) [134, 135, 136].

The 3-bit scale factor determined for 32 words must of course be transmitted together with the transmission of parity bits, the signalling-in-parity coding method being used with the word at the coder output having a length of 11 bits.

After implementing data compression using the scale factor, the frame length is $64 \times 11 = 704$ bits. The 24 control bits which are needed by the

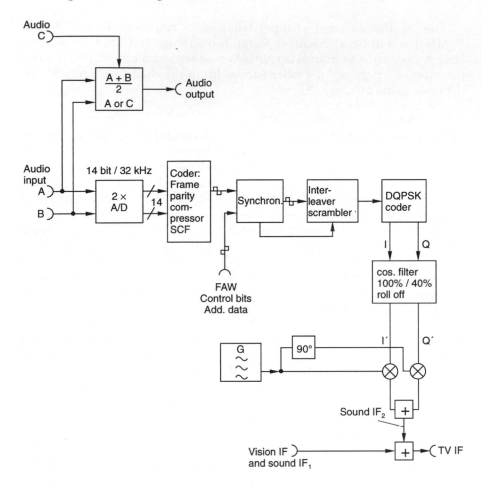

Fig. 3.18 NICAM coder/modulator

receiver for frame alignment, operating mode identification and decoding of the audio data, are added ahead of these frames. The control bits C1 to C4 for specifying the operating mode are defined in Table 3.3. The control bits allow further features to be added for future applications. Fig. 3.19 shows the structure of a NICAM 728-bit stereo packet. In the interleaver and scrambler the 704-bit packets from the coder are interleaved and scrambled, providing a pseudo random sequence. This measure is to reduce burst errors and simplify the derivation of the transmission clock in the NICAM receiver.

At the input of the modulation section of the NICAM transmitter, the data are coded using DQPSK. The coder combines two bits at the input to form a bit pair corresponding to a four-valued symbol. Prior to QPSK modulation the inphase and quadrature signal components are subjected to

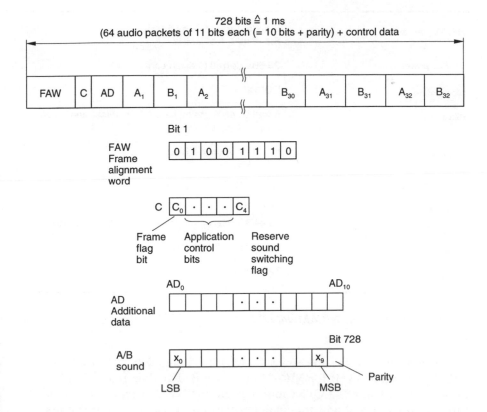

Fig. 3.19 Structure of a NICAM 728-bit frame with stereo signal (prior to interleaving)

C1	C2	C3	C4	Operating mode
0	0	0		Stereo
0	1	0		Two independent mono channels
1	0	0		One mono signal and one data channel
1	1	0		Pure data transmission (no audio signal)
			1	NICAM mono sound signal modulated onto FM carrier
			0	FM signal other than NICAM signal

Table 3.3 NICAM operating modes

NICAM carrier	Standard B and G: 5.85 MHz + f_{vision} Standard I: 6.552 MHz + f_{vision}
Carrier power	-20 dB referred to vision carrier peak
Modulation	DQPSK
Filter	B and G: 40% cosine rolloff (transmitter and receiver) I: 100% cosine rolloff
Bandwidth	B and G: approx. 510 kHz I: 728 kHz
FM carrier power	-13 dB (B, G); -10 dB (I)
Data rate	728 kbit/s
Sampling	14 bit/32 kHz, compressed to 10 bits per sample
Error protection	1 parity for 10-bit sample
Block format	728 bits per 1 ms with 8-bit FAW

Table 3.4 Characteristic NICAM data

cosine filtering for the necessary bandwidth, being 728 kHz for NICAM Standard I and 510 kHz for NICAM Standard B. The data rate is 728 Kbit/s in both cases. Following QPSK modulation, the NICAM signal is combined with the video signal and the analogue FM mono carrier for transmission (Table 3.4) [132].

3.3.3.2 Decoding
The essential task of the NICAM decoder is to reconstruct the NICAM datastream, retrieve the clock for the D/A convertor, descramble and deinterleave and restore the original audio data in line with the scale factors.

A multistandard receiver of analogue design suitable for NICAM sound signals and FM analogue sound signals would need a large number of different filters, FM demodulators as well as a NICAM decoder meeting the above requirements. To manufacture such an analogue multistandard receiver would not be economically viable but a multistandard audio chip set with four integrated circuits fulfilling the above requirements is available from ITT Intermetall (Fig. 3.20) [132].

The MSP (multistandard audio processor) in conjunction with the AMU (audio mixer unit) and ACP (audio processor) provides complete audio processing. The data are transferred via the digital S-bus which is a serial bus with the clock, ident and data lines. On the data line, four 16-bit channels are transferred in time multiplex. The central CCU (microcontroller unit) controls the system ICs via the digital control bus, i.e. the IM-bus.

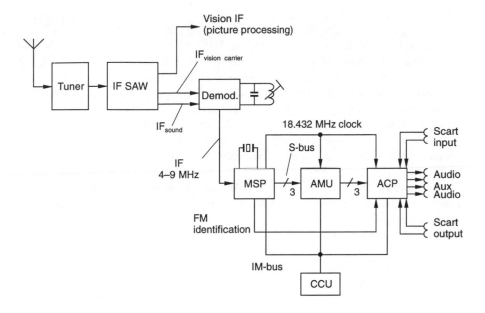

Fig. 3.20 Multistandard audio receiver fitted with ITT ICs

The high performance of these ICs is demonstrated by way of the Multistandard Audio Processor MSP 2400 which is a CMOS IC with integrated A/D convertor and RISC processor and demodulator for FM/AM and DQPSK using programmable frequency mixers (range 0 to 9 MHz). The chip gives digital demodulation and decoding of NICAM coded TV sound signals and FM/AM modulated mono sound signals. Moreover, it is also able to demodulate FM signals according to the analogue dual-carrier method. The chip can be adapted to the different carrier frequencies of the various transmission standards by using appropriate mixer and filter coefficients while able to decode signals of the two NICAM versions for standards B and I. A special pay-TV mode can be implemented in the NICAM mode. The analogue sound IF signal is applied to the input of the chip which provides up to three AF baseband sound channels on the S-bus. Baseband processing including de-emphasis filtering and subsequent D/A conversion are performed in the AMU and ACP.

3.3.3.3 Measurements

For the operation of NICAM systems (and development and production of NICAM components) it is necessary to have TV RF and IF signals containing a NICAM sound carrier in addition to the vision and FM sound carrier. The TV RF test transmitters must also be fitted with NICAM generators delivering a complete serial 728 Kbit/s datastream with frame alignment word, control and additional data bits plus the digital information of different audio-frequency sinewave signals required for testing NICAM demodulators.

To check the dynamic range of the NICAM sound system, the TV test transmitter with NICAM generator produces a low-frequency sinewave signal for driving the analogue/digital convertors. The quantisation noise of the convertors primarily determines the dynamic range of digital systems. The low-frequency sinewave signal makes the convertor in the NICAM receiver analyse its characteristics, to determine the dynamic range. After eliminating the low-frequency signals with a highpass filter that follows the D/A convertor, the dynamic range of the NICAM receiver can be determined [133].

To increase the data capacity, the NICAM system provides error detection merely by means of parity bits. Bit errors can be concealed but not corrected. Even moderate bit error rates cause noise that disturbs the listener and necessitates automatic switchover of the NICAM receiver to the FM sound carrier in mono mode. To check the switchover function of NICAM receivers, the TV test transmitter with NICAM generator must be able to produce signals with defined bit error rates.

For measuring the NICAM signal delivered by a TV transmitter, a NICAM demodulator is used which, compared with the NICAM-compatible consumer receiver, features optimum quality of signal processing plus a variety of indicators and outputs for the digital and analogue signals enabling monitoring of the transmitted NICAM signal and the analysis of errors. To analyse, for example, the transmission reliability, the I and Q components can be applied to an oscilloscope to display the vector diagram for the QPSK signal and the eye pattern. Evaluation of eye height, eye width and trace density facilitates error detection.

The actual bit error rate is obtained by monitoring the six most significant bits of the 10-bit data words. The variance of these bits permits the complete transmission spectrum to be checked.

3.4 Analogue TV colour systems

3.4.1 PALplus

The television standards PAL, NTSC and SECAM were introduced more than 25 years ago. They provide the viewer with a 625 line colour TV picture of relatively good and accepted quality (525 lines with NTSC). Measured by today's standards, these colour systems show various weaknesses. With the PAL system, these are in particular the crosstalk of the luminance signal components into the chrominance channel (cross chrominance or cross colour) and the crosstalk of the chrominance signal into the luminance channel (cross luminance). The cause of these effects is frequency multiplexing. These cross effects occur in particular when the filtering in the home receiver is inadequate. Digital signal processing today allows reduction of cross chrominance and cross luminance by adaptive comb filter decoders in more complex PAL decoders [138].

The separation of the chrominance and luminance spectral components in the receiver is done in such a way that the cross effects are not humanly visible — an idea used by the compatible I-PAL and Q-PAL methods. With both these methods, the luminance channel is divided into a low-frequency and a high-frequency channel. The high-frequency luminance channel and the entire chrominance channel are transmitted in alternate lines. The difference between I- and Q-PAL lies in the complexity of filters used in the receiver for separating the spectral components [139].

The PALplus system is a further step towards compatibility with the new 16:9 aspect ratio. PALplus is an extended PAL standard specified during a three-year research project started in 1990 and jointly supported by European public broadcasters and major consumer equipment manufacturers. The introduction of this system in Germany for instance was scheduled for 1994.

The changeover from a TV picture aspect ratio of 4:3 (width:height) to the new 16:9 ratio better represents the field of human vision. The 16:9 aspect ratio takes into account that the visual range seen by the eye is greater in the horizontal plane than in the vertical. Moreover, 16:9 (1.77:1) is a compromise with the aspect ratios used in the cinema. European wide-screen films have an aspect ratio of 1.66:1 and cinemascope films 2.35:1 [141]. Therefore, wide-screen films will in future no longer be viewed with the black bands top and bottom on the home receiver.

Although the new TV standards D2-MAC, HD-MAC and HDTV (Sections 3.4.2 and 3.4.3) feature the 16:9 aspect ratio, they cannot transmit in the terrestrial 7 MHz channel (VHF) and 8 MHz channel (UHF). They are initially intended for distribution via satellites and cable channels. This means further innovation for the PALplus system, since homes with terrestrial TV reception would not be able to receive 16:9 programmes. A separation of the satellite and cable services from terrestrial television and hence splitting of TV viewers into those receiving 16:9 or only 4:3 programmes, would have economic and technical disadvantages, since the 16:9 programmes would also have to be transmitted in the 4:3 format. In addition to cable and satellite, comprehensive terrestrial TV broadcasting will continue to be the mainstay of TV coverage, and so the new systems must be distributable through all existing TV media.

The main aim of PALplus is therefore to provide a better signal quality than PAL, a 16:9 aspect ratio fully compatible with existing 4:3 receivers, and improved sound quality [139].

3.4.1.1 Format conversion
The reproduction of a 4:3 picture on a 16:9 receiver (Fig. 3.21) is a straightforward process. The black bands on the 16:9 display on the right or

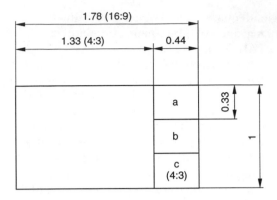

Fig. 3.21 Representation of 4:3 picture on 16:9 display with additional preview pictures a, b, c

left side or both sides could be used for a picture-in-picture feature. With the 16:9 receivers, the overall picture would consist of the main picture and inset pictures controlled by a zoom function and joystick.

Compatible display of 16:9 pictures on 4:3 monitors is more critical. Basically, there are four possibilities of solving this problem (Fig. 3.22) [139]:

- The simplest way is the so-called side-panel or edge-crop method, which is mainly used in European countries. The 16:9 picture is adapted to the 4:3 ratio by suppressing a side panel on the left and right side of the 16:9 picture, assuming that these side panels do not contain picture-relevant information.
- With the letterbox method the 16:9 picture is transformed into the 4:3 picture format causing a black band to appear at the top and at the bottom of the picture. Viewers in Europe are already accustomed to this format by the transmission of cinemascope films.
- The above two approaches are combined in a compromise, where a narrow side panel is eliminated and a narrow black band is present at the top and bottom.
- With the fourth method, the 16:9 picture is horizontally compressed in time and forced into the 4:3 ratio. Circles become elliptical with the longitudinal axis in the vertical direction. The resulting picture is distorted, but this squeeze method is being discussed and used.

3.4.1.2 Letterbox method
With the letterbox method, the 16:9 picture produced in the studio is subjected to vertical decimation filtering to reduce the height of the picture. For this process, 576 active lines are assumed, i.e. 625 lines minus twice the field blanking interval of 24.5 lines. The active picture is reduced to 432

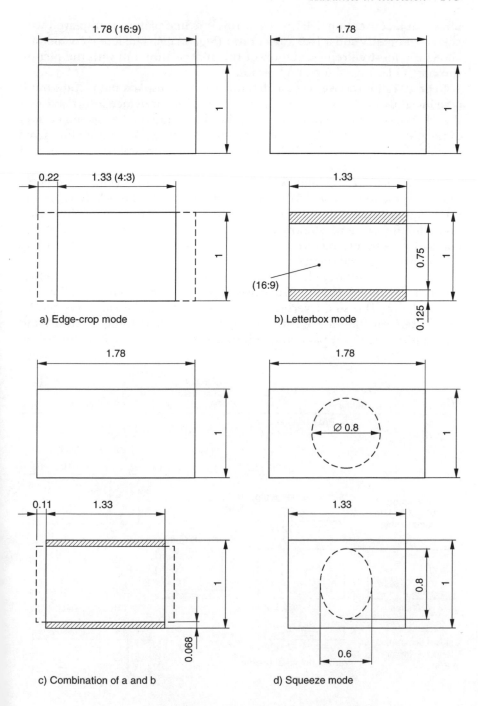

Fig. 3.22 Possibilities of representing 16:9 picture on 4:3 display

lines. On a conventional 4:3 receiver this 432 line picture is displayed with correct geometry and a 16:9 aspect ratio (Fig. 3.22*b*). The letterbox method allows the loss-free representation of the 16:9 picture, but with the picture showing a black band top and bottom.

The PALplus receiver with a 16:9 picture tube displays the picture but it is reduced to 432 lines from 576 active lines. The luminance information of the 144 removed lines is transmitted in the black bands at the top and bottom of the picture. In the 72 lines at the top and 72 lines at the bottom the signal is transmitted in the vertical helpers for it to remain invisible on the standard 4:3 receiver (Fig. 3.23).

The vertical information (vertical helpers) is derived from a reversible line interpolation (Table 3.5) [138]. From the original lines A, B, C . . . the information reduced to 432 lines is derived by interpolation of neighbouring lines. In the example shown, the lines 'A, B, C' are transmitted in the compatible 432 line picture and line C in the two vertical helpers. Lines 'A,B,C' are displayed on the 4:3 receiver.

In the 16:9 receiver the information contained in the vertical helper signals is processed and the 576 active lines of the 16:9 picture retrieved by extrapolation. Due to the vertical decimation filtering, impairments may be caused but can be reduced by an improved method using vertical band segregation into lowpass and highpass components. For 'invisible'

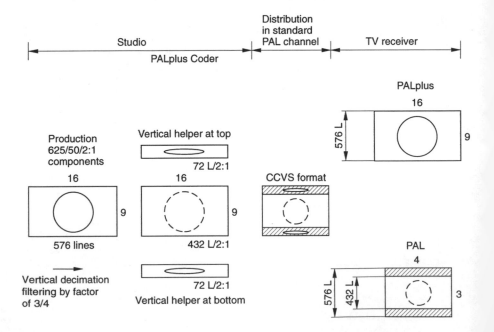

Fig. 3.23 Conversion and transmission of PALplus picture

Source	Transmission		Sink	
16:9 picture 576 lines	432 lines	2 x 72 helper lines	4:3 display	16:9 display
A	A' = A		A'	A
B				B = 3/2 B' - 1/2 C (c)
	B' = 2/3 B + 1/3 C (a)		B'	
C		C		C
	C' = 1/3 C + 2/3 D (b)		C'	
D				D = 3/2 C' - 1/2 C (d)
E				

c by solving a to B
d by solving b to D

Table 3.5 Line interpolation in PALplus letterbox method [138]

transmission of the vertical helpers the luminance information — after filtering and nonlinear precorrection — is modulated onto the colour subcarrier by vestigial sideband amplitude modulation. The amplitude precorrection makes the signal less susceptible to noise and echo effects in the channel. Due to the colour subcarrier modulation, the resulting spectrum of the helper is freed from low-frequency signal components which may cause sync. errors especially in older type receivers. Moreover, the visibility of the helper signal on the 4:3 receiver is reduced (Fig. 3.24).

The PALplus specification is also aimed at increasing the useful luminance bandwidth to obtain on the 16:9 screen the same horizontal resolution as on the 4:3 screen. If the aspect ratio 16:9 and 4:3 is referred to the same picture height, the picture area for the 16:9 aspect ratio is increased by the factor (16:9):(4:3)= 1.33. According to the signal theory, a bandwidth that is increased by this factor would be required to display the two pictures with the same resolution. For a PALplus signal, this would mean a video bandwidth of 5 MHz × 1.33 = 6.67 MHz, which is not implementable with the 7 MHz channel in the VHF range.

There is, however, a certain bandwidth reserve. Due to the colour subcarrier trap in 4:3 receivers, the luminance bandwidth in standard 4:3 receivers is reduced to 3.5 to 4 MHz. Assuming an average value of approximately 3.75 MHz, a minimum bandwidth of 3.75 MHz × 1.33 = 4.98 MHz is obtained for PALplus with comparable horizontal resolution of

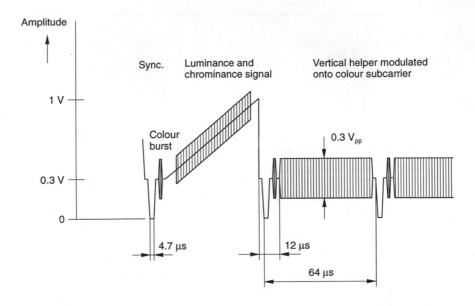

Fig. 3.24 Representation of CCVS with one picture line and one helper line

the luminance signal. With the aid of an appropriate PALplus decoder making optimum use of the transmission channel capacity, a 16:9 picture can be transmitted with the sharpness one is accustomed to from the standard 4:3 transmission [141].

3.4.1.3 PALplus coder and decoder

Fig. 3.25 shows the block diagram of the PALplus coder. In terms of picture frequency, number of lines and line duration, the received 16:9 picture has the same system parameters as the conventional 4:3 picture. Because of the similar line duration, the horizontal picture on the compatible receiver is compressed by a factor of $3/4$, resulting in a vertical picture decimation by the same factor. The 576 active lines are subjected to a vertical scan conversion to yield a main picture of 432 lines and 2×72 lines. The scan conversion is implemented by vertical band segregation into a lowpass and a highpass component. The lowpass component is transmitted with an amplitude of 0.3 V in the black and blacker-than-black range, i.e. symmetrically about the blanking value ± 0.15 V. After preprocessing, the main signal and the helper information are correctly placed in the multiplexer. Colour coding for the main picture is made by a special colour coder so that the horizontal bandwidth of 5 MHz will be preserved for the luminance signal for vertical structures.

The PALplus signal is displayed by 4:3 receivers without an additional decoder. The improved luminance-chrominance coding reduces the cross

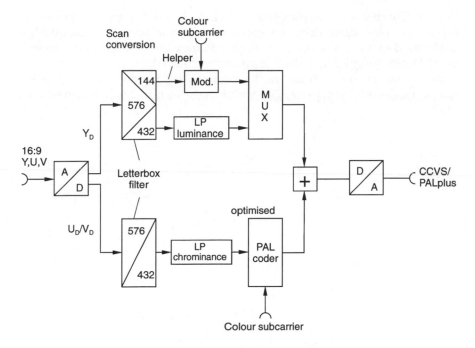

Fig. 3.25 PALplus coder (in camera mode)

effects for the 4:3 receiver too. In the 16:9 receiver a colour decoder matched to the coder ensures optimum luminance-chrominance separation. After the colour decoder the signal is routed via a demultiplexer with picture memory. The signal is split into the highpass and lowpass component. The scan is then converted from the 432 and 144 lines to the original 576 lines (Fig. 3.26).

The principle of vertical band segregation is based on splitting the input signal into a lowpass and a highpass component with interpolation/decimation to 432 and 144 lines. The filters used are of great importance since they feature a high stopband attenuation. After the decimation, the spectra will overlap at the band limits, causing aliasing in the 4:3 receiver. If the cutoff frequency were selected too low, a band gap would occur in the 16:9 receiver following band synthesis. Aliasing in the highpass and lowpass components after decimation cannot be avoided in practice. In the 16:9 receiver, aliasing components are mainly compensated by QMF(quadrature mirror filter). Despite the overlapping ranges, the QMF allows almost error free reconstruction of amplitude and phase at the band limits [140, 141].

The PALplus project is also aimed at improving sound quality. The problem is the compatibility with the sound systems used in other parts of

Europe. This is a very complex situation since in addition to different sound carrier spacings different-stereo systems are also in use: the analogue IRT dual-sound method and the digital NICAM method. The solution can only be a PALplus sound signal which is interleaved into the spectrum in addition to the sound transmission system. To minimise the bandwidth, a high-grade data reduction method, preferably MUSICAM (Section 2.4.2), has to be used [137].

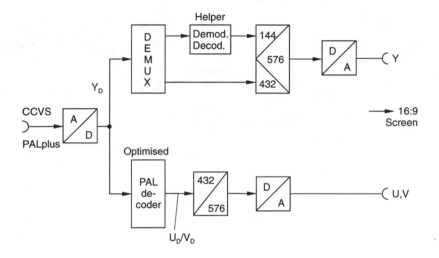

Fig. 3.26 PALplus decoder (in camera mode)

Another aspect to consider with PALplus is that, for reasons of cost, foreign-language films in Europe are often not dubbed, but broadcast in the original language with subtitles. The text is inserted at the bottom of the picture, in the black band in wide-screen films. In the letterbox mode this is, however, not possible because of the helper lines. An insertion into the active letterbox picture would however reduce the viewing area.

3.4.1.4 Outlook
The PALplus system simulation phase is almost completed. The first public demonstration of PALplus was at the IBC 1992 (International Broadcasting Convention) in Amsterdam. PALplus receivers containing special ICs were to be available by 1995. Following trial transmissions from autumn 1994, regular operation of PALplus would have started at the International Consumer Electronics Show 1995 in Berlin and according to the time schedule, the complete infrastructure for the distribution of 16:9 pictures in the terrestrial transmitter network would follow. In view of the high degree of compatibility, time is not really a crucial factor. Standard PAL broadcasts

and PALplus broadcasts can be combined as desired during the ongoing transition. Signalling of the PALplus transmission provides for automatic conversion in the PALplus receiver. For the complete range of programmes offered, the new aspect ratio will be gradually introduced until the 16:9 ratio is accepted as the industry standard [137].

3.4.2 D2-MAC

In the D2-MAC system the video signal consisting of the luminance and the chrominance signal is transmitted sequentially. In addition to the Y signal and the U/V signal a data burst for the digital sound and auxiliary data is inserted into the television line. This means that the D2-MAC system combines digital and analogue techniques. One reason for this combination is certainly the fact that at the time of standardisation video baseband coding techniques and technology had not yet reached the required standard for a bit rate of 20 to 40 Mbit/s. On the other hand, this means that the D2-MAC evolution path will inevitably come to an end as soon as appropriate picture coders are available.

The D2-MAC/packet system has been specified for direct broadcasting satellites. It was derived from the C-MAC/packet system, which according to the resolutions of the EBU of April 1985 should be used wherever a large capacity of sound and data signals and optimum picture quality (when changing over to the wide aspect ratio 16:9) are the dominant considerations. Accordingly, D2-MAC/packet should be used on satellite links if the signals are to be fed into the cable networks comprising 7 MHz or 8 MHz channels without being recoded. This means that D-MAC and D2-MAC are the cable version of the C-MAC system, which requires cable channels of at least 10.5 MHz. D2-MAC is nevertheless intended for distribution in cable networks in the 12 MHz hyperband channels. D2-MAC/packet is thus suitable for all media including terrestrial distribution. In France especially, the terrestrial use of D2-MAC is being considered, giving a chance to replace SECAM. Moreover, D2-MAC is a European system, avoiding the drawbacks of the different PAL and SECAM systems in international programme exchange [147].

3.4.2.1 D2-MAC elements
The acronym D2-MAC/packet is made up of the following elements:

- D2 stands for sound and data transmission during the time window with duobinary modulation at half the bit rate of C-MAC which is able to transmit 20.25 Mbit/s.
- MAC stands for multiplex analogue components and means the analogue transmission of the Y and U/V signal components in time-division multiplex (Fig. 3.27). In contrast to the PAL system, where the luminance and the chrominance signal are transmitted in· frequency multiplex

causing cross luminance and cross colour effects, the Y and U/V components are transmitted in time-division multiplex in the MAC system to avoid these crosseffects.

- The word 'packet' means that the digital sound and data signals are transmitted in packet form. The digital information of each packet is provided with a header, and auxiliary information such as identification of the broadcaster, the original language or TV programme type (news, sport, music) can be transmitted in addition to the digital sound signals.

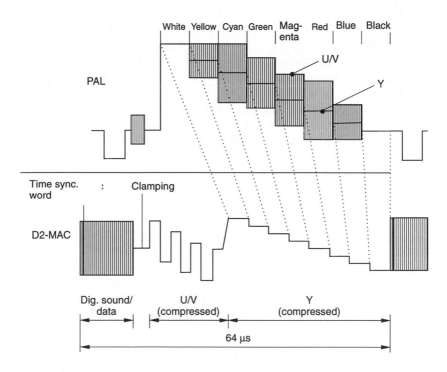

Fig. 3.27 Colour bars with D2-MAC/packet signal and with PAL signal

The time-division multiplex of a D2-MAC line provides the following sequence: data signals for the sound channels with sync. bits and identification etc., analogue colour information within alternate lines (colour difference signal U in odd-numbered lines and V in even-numbered lines), followed by Y luminance signal [142].

Due to the time-sequential transmission of luminance and chrominance in a television line of 64 μs, the analogue components for luminance have to be compressed by a factor of 3:2 and for the chrominance signal by a factor of 3:1. The clock rate of the system of 20.25 MHz is derived from these compression factors, the resulting number of clocks per line being 1296 (Fig. 3.28).

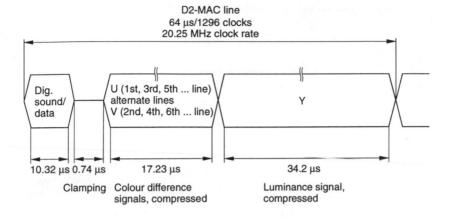

D2-MAC line
64 µs/1296 clocks
20.25 MHz clock rate

Dig. sound/data

U (1st, 3rd, 5th ... line) alternate lines
V (2nd, 4th, 6th ... line)

Y

10.32 µs | 0.74 µs | 17.23 µs | 34.2 µs

Clamping | Colour difference signals, compressed | Luminance signal, compressed

Fig. 3.28 D2-MAC baseband signal

The Y and U/V signals are compressed in the D2-MAC coder, in which after analogue to digital conversion the signals are separately buffered and read out at a correspondingly higher clock rate. The subsequent digital to analogue conversion provides the compressed analogue signals for combining with the data bursts. Expanding in the decoder of the receiver is of course reciprocal.

3.4.2.2 Sound/data transmission
In the C-MAC/packet system the databurst uses a bandwidth-saving PSK modulation and therefore has a bit rate of 20.25 Mbit/s in the transmission window. The 20.25 MHz clock rate is common to all MAC versions being derived from the 13.5 MHz sampling frequency of the 4:2:2 studio standard for the luminance signal (Section 3.5.1), multiplied by the compression factor 1.5. For the chrominance signal the clock rate is the sampling frequency of 6.75 MHz multiplied by the chrominance compression factor 3, yielding 20.25 MHz.

In the D-MAC/packet system the PSK used in C-MAC is transformed into a duobinary digital signal, the bit rate of 20.25 Mbit/s being retained. Duobinary coding uses three logic levels (Fig. 3.29) meaning that in contrast to the binary code it is a ternary coding method. The binary input signal is first transformed into a signal in which the level remains constant when the input signal is at logic 1 and the level changes when the input signal is at logic 0. The duobinary signal is derived by averaging between two neighbouring individual values. This results in a smaller bandwidth of the duobinary signal which has a zero crossing at exactly half the clock frequency of 10.125 MHz [143].

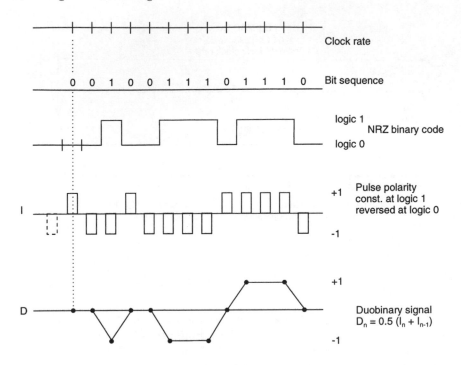

Fig. 3.29 Duobinary coding of D2-MAC data (bandwidth reduction by about 30%)

The duobinary modulation has a resistance to reflections, especially in cable distribution networks.

The databurst comprises 82 packets of 751 bits each at the beginning of the first to the 623rd line of the D2-MAC signal (Fig. 3.30). The transmission reliability is ensured by special error protection codes and error protection bits (bit interleaving, parity bits, Hamming code).

For sound transmission, the D2-MAC system provides channels with a bandwidth of 15 kHz and, for instance, commentary channels of 7 kHz. The high-quality channels use 14-bit linear coding companding to 10 bits and high error protection by the Hamming code. The medium-quality channels provide the 7 kHz commentary sound with 14 bits companded to 11 bits and with low error protection by the parity bits. Companding is similar to the NICAM method (Section 3.3.3). The Hamming code uses five protection bits per sample., the low error protection being a simple parity check with one parity bit per sample.

The sound and data transmission capacity of D2-MAC, not in the transmission window but linear in time, is 1540 Kbit/s in total, and derived from the window bit rate of 10.124 Mbit/s and the duration of the databurst of 10.32 μs relative to the line duration of 64 μs, less the overhead bits [144, 145, 146].

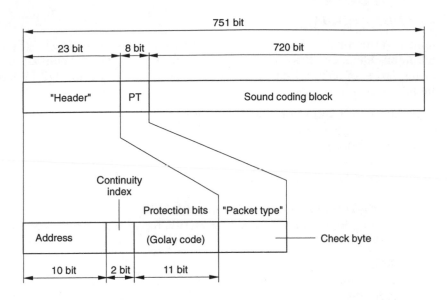

Fig. 3.30 Digital sound transmission with D2-MAC packet

3.4.2.3 Summary of D2-MAC
The advantages of D2-MAC are:

- Baseband signal is complete with video and sound, a special advantage for transmission in line networks.
- D2-MAC has no sync. pulse and just one FM carrier for satellite transmission, whereas PAL has four carriers (vision carrier, colour subcarrier, two sound carriers). This allows a greater frequency deviation for D2-MAC and in conjunction with the reliable digital sound and data transmission the use of a smaller diameter receiving antenna.
- D2-MAC is a European system, which provides a high degree of flexibility also with a view to future new services due to signal transmission in time-division multiplex and digital sound and data transmission.
- The system is able to transmit four companded high-quality sound channels (two stereo or four mono channels) plus 190 Kbit/s for data.

  Disadvantages of D2-MAC are:
- The reduction of the vertical resolution of chrominance from 2.5 'MHz' of PAL to 1.4 'MHz'.
- The time compression reduces the luminance S/N ratio by 2 to 3 dB as compared to PAL [143].

3.4.3 HDTV/HD-MAC

Similar to the quality of sound gained from the compact disc, HDTV (high definition television) sets a standard, which in picture quality is comparable to the clarity of 35 mm films. Compared to existing TV systems, HDTV has the following advantages:

- double resolution in horizontal and vertical direction
- flickerfree reproduction
- improved colour reproduction
- aspect ratio of 16:9
- multichannel sound in CD quality.

3.4.3.1 Analogue HDTV methods in Japan and the USA

Japan was the pioneer in the subject of HDTV, for which a studio standard was defined in 1970:

- 1125 lines
- line interlacing
- 5:3 aspect ratio
- 60 Hz field frequency.

The studio standard was the basis for the transmission standard for the MUSE system. According to MUSE, only changing pixels are transmitted. The transmission capacity remains constant, with the slight change between two consecutive pictures used to increase the picture resolution. If there is great movement in the picture only coarse changes can be transmitted, but this goes unnoticed by the human eye.

This concept, although not compatible with existing TV receivers, was experimentally tried using the Japanese satellite BS-2B in 1989 and consequent HDTV/MUSE programmes were operational in 1991 [139].

In the USA the Japanese studio standard with 1125 lines/60 Hz has largely been accepted. However, there is the problem of terrestrial TV channels having a bandwidth of 6 MHz and priority being given to terrestrial programme distribution over other transmission media. Analogue HDTV certainly cannot be implemented in the 6 MHz channel, therefore, the US were the first country to enter, and are now leading, the field of terrestrial digital HDTV.

3.4.3.2 HD-MAC overview

The European HD-MAC is based on the previous MAC system for satellite broadcasting and was planned to be operational by 1995. Since the information transmitted with HD-MAC is four times the quantity compared to PAL (twice the number of lines and twice the number of frames), a bandwidth of approximately 30 MHz would be required. By present day standards, this bandwidth could only be achieved with broadband transponders (36 or 70 MHz) or in fibre optic distribution networks. The

HD-MAC system therefore uses a data reduction in the sense of a motion-adaptive picture processing [1, 149]. Over small areas, motion vectors are derived from a few pixels and transmitted as additional picture information in the digital data channel of the HD-MAC signal. The HD-MAC receiver contains a motion processor for reconstruction of the pictures. Moving pictures are subdivided into three categories: little, normal and fast motion. With little motion, a picture with 16:9 aspect ratio and 1250 lines is produced every 80 ms. With a field frequency of 50 Hz, a picture period of 20 ms is obtained, which means that in this case a high-definition picture is composed of four fields. With normal motion, half the resolution is used, with a picture being produced every 40 ms, meaning that two fields make up a picture. With fast moving picture sequences, a picture is transmitted every 20 ms, so that a spatial resolution corresponding to 625 lines is achieved. Due to the slowness of the human eye in recognising picture sharpness in fast moving picture sequences, there is no subjective loss in quality. The motion-adaptive picture processing leads to a bandwidth reduction to 12 MHz. The status of the picture processor depending on the 80, 40 or 20 ms mode is transmitted via a DATV (digitally assisted television) signal. A data signal with duobinary coding is transmitted at 20.25 Mbit/s in the field blanking interval of the HD-MAC signal [153]. This data signal contains a branch decision BD (80, 40 or 20 s process), and in the case of the 40 ms mode decision a motion vector MV.

HD-MAC transmission starts with the HDTV studio signal, which is coded into a video signal and an additional digital control signal in the field blanking interval. The video signal contains the luminance and chrominance signals after bandwidth reduction according to the MAC/packet system. Normal-definition (ND) receivers are able to display a MAC-like signal with little impairments. The digital control signal DATV is specially transmitted for the HD-MAC receiver. It conveys the information required for reconstructing the HDTV signal from the reduced bandwidth compatible signal. This DA data signal allows the HD-MAC receiver to act as a slave to the transmitter coder, so that reliable motion-adaptive video processing can be performed in the decoder. The maximum bit rate of the DA signal is about 1 Mbit/s.

3.4.3.3 HD-MAC coding principles
The HD picture has a spatial resolution of twice that of an ND picture in both the horizontal and vertical directions. To satisfy the requirement for compatibility, the bandwidth has to be reduced by a factor of four. The HD-MAC coding system uses an adaptive spatio-temporal subsampling pattern for bandwidth reduction. In a block of 16 HD lines by 16 HD pixels the coding system distinguishes between three modes as a function of the picture contents:

- The 'stationary' 80 ms mode is associated with very slow movements up to 0.5 pixels per 40 ms. The spatio-temporal subsampling used in this mode is a frame-line quincunx pattern with a time interval of 80 ms (Fig. 3.31) (quincunx: pattern of 5 sampling points). The reconstruction of the HD picture is achieved by an appropriate interpolating filter. This mode yields the highest spatial resolution for slowly moving picture components.
- The 'tracking' 40 ms mode is associated with the medium fast movements with up to 12 pixels per 40 ms. The spatio-temporal subsampling pattern used in this mode is a field-line quincunx applied to the odd fields. For each block of 16 by 16 pixels in the even fields a motion vector *D* is determined. These fields can be approximated from the adjacent fields using a symmetrical filter applied in the direction of the vector *D*. The HD picture is therefore reconstructed by means of a spatial interpolating filter on the odd fields, followed by a motion-compensated temporal interpolation of the even fields. The spatial resolution in this mode is half of the 80 ms mode. The efficiency of the 40 ms mode essentially depends on the performance of the motion estimation algorithm.
- The 'dynamic' 20 ms mode is associated with the fast movements. A pure spatial subsampling pattern, which is a field-line quincunx with a horizontal period equal to four HD pixels, is used in this mode. In the receiver the HD picture is reconstructed by means of a spatial interpolating filter, the resolution being reduced to one fourth of that of the 80 ms mode.

As shown in Fig. 3.32, the three modes 80, 40 and 20 ms process the input HD picture independently of each other. Their outputs are selected as a function of the motion velocity ('*a priori*' decision) and the comparison between the vector-corrected reconstructed picture and the original, i.e. the reference picture ('*a posteriori*' decision).

The *a posteriori* decision is made by comparing the spatially prefiltered (in all modes) and temporally interpolated pictures (in 80 ms and 40 ms modes) with the original pictures. The mode which yields the smallest difference energy is selected. This decision procedure can be ambiguous when modes perform very closely to each other. This ambiguity can be eliminated by means of an *a priori* decision, which uses a motion vector to check whether picture contents are moving faster than 12 pixels per 40 ms. In this case, the decision is forced to the dynamic 20 ms mode irrespective of the *a posteriori* decision.

Another decision criterion is derived from a consistency check over a wide spatio-temporal range. For instance, it is not very useful to have one block in a completely different mode to that of its surrounding neighbours. This exotic block may cause noticeable switching artefacts. Consistency in the temporal domain is also very important since there will be little coding gain from a block which, for instance, remains in the stationary mode for only 80 ms [150, 151, 152].

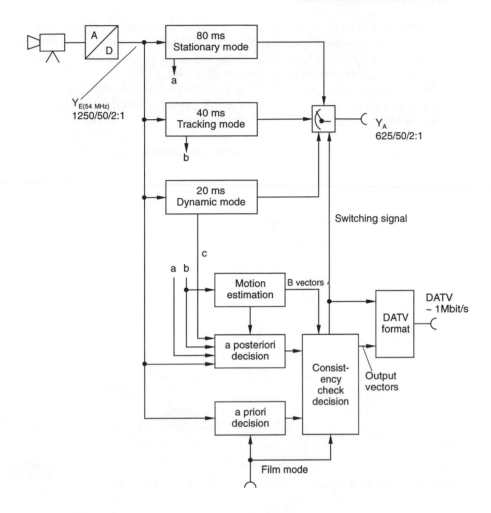

Fig. 3.31 Luminance bandwidth reduction in HD-MAC coder

The HD chrominance signals are also reduced in bandwidth by using sub-Nyquist sampling. Motion compensation is not used for the chrominance signals. The mode decisions in the 80 ms and 20 ms modes are the same as the luminance decisions. In case of the 40 ms mode in luminance, there are two possible modes in chrominance: if the motion vector is less than or equal to, for example, six samples per 40 ms, the chrominance mode is also 40 ms. For larger motion vectors the 20 ms mode will be used for chrominance.

To improve compatibility, the following measures are required:

- In the 80 ms mode the ND-MAC picture may show disturbing edge crawling due to the potential high frequency content. This condition has a temporal periodicity of 12.5 Hz. In order to eliminate this condition, a vertical lowpass filter is included in the HD-MAC coder. In the HD-MAC decoder reciprocal highpass filtering has to be employed which has the disadvantage of enhancing transmission noise.

• Due to the motion compensation, the compatible ND-MAC picture may exhibit a 25 Hz judder effect. The judder is introduced by two successive fields originating from the same movement. An improvement is achieved by temporal filtering in the coder and in the decoder, mixing the samples to be transmitted with the compensated samples in the direction of motion (line shuffling).

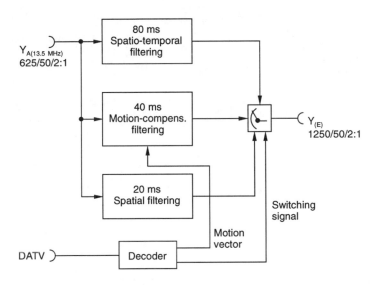

Fig. 3.32 HD-MAC decoder for luminance bandwidth reduction

3.4.3.4 Data packet in HD-MAC signal

In addition to the MAC compatible picture signals a data packet with duobinary coding is transmitted in the HD-MAC signal. The packet structure corresponds to that defined in the MAC/packet specification. One packet contains a block of 751 bits, which are divided into the header and the data area. The header contains the address code, which is allocated to a particular broadcast service, e.g. a sound signal or the DATV data signal. The data area has a length of 91 bytes, the first byte being used as a packet type byte to identify the use of the packet data.

The HD-MAC sound is coded according to the MAC/packet specification to ensure compatible reception. The sampling frequency is 32 kHz for high quality and 16 kHz for medium quality. The coding modes are linear coding with 14 bits/sample or near instantaneous coding with 10 bits/sample. Simple error protection uses 1 parity bit/sample and high-grade protection a 5-bit Hamming code per sample.

HD-MAC is downward compatible with D2-MAC, enabling HD-MAC to be distributed through the existing MAC transmission channel and be reproduced with 625 lines on a D/D2-MAC receiver with reduced resolution compared to the source signal. For transmitting the HD-MAC signal via satellite it is intended to use a channel bandwidth of 36 MHz instead of the 27 MHz usual for DBS. Through medium-power satellites the signal can also be received by systems with 60 cm dishes. The HD-MAC signal can also be transmitted in the 12 MHz hyperband in broadband cable networks. This is made possible by the use of steep-edged filters and vestigial sideband amplitude modulation of the vision carrier.

HD-MAC has been designed as a new service for high quality TV and to be compatible with the existing MAC/packet services. The European HDTV specification as defined in the EUREKA 95 project was therefore based on the MAC/packet family standard laid down in the CCIR-Reports 10-1073 and 1074 [148].

3.5 Digital TV studios

The TV studio technique differs fundamentally from the transmission technique. In the studio, postprocessing is a basic requirement where the TV signal must be of a high fidelity nature even in the case of multiple postprocessing to avoid a noticeable loss in quality.

In new TV studios, a digital infrastructure, component coding and an aspect ratio of 4:3 and 16:9 are usually found today [154].

3.5.1 Studio standard

CCIR Recommendations 601 and 656 (Tables 3.6 and 3.7) are used as a standard for the introduction of digital component studio techniques.

Recommendation 601 describes in detail the sampling parameters of the 4:2:2 studio standard and provides a hierarchical approach in establishing standards with a view to resolution and compatibility.

For higher members of the studio component family the sampling frequencies of the luminance signal and the two chrominance signals (or, if used, the red, green and blue signal) can be related by the ratio 4:4:4. A preliminary specification of the 4:4:4 member is included in CCIR Recommendation 601 Annex I [157].

Recommendation 656 defines the interface standard (coding of data signal) and the studio transmission standard [158]. Following the initial standardisation, there have been some extensions: the word length has been

Sampling frequency	Y		13.5 MHz		
	C(R-Y)		6.75 MHz		
	C(B-Y)		6.75 MHz		
Number of samples per line (625-line system)	Y 864 C 432		Component Active line	Y 720 C 360	
Sampling structure	Orthogonal				
Resolution	8 bit PCM (10 bit PCM)				
Quantisation with scale 0 to 255	Black level		16		
	White level		235		
	Chrominance		128 + 112		
Reserved codewords	0.255				

Table 3.6 Signal parameters of CCIR Rec. 601

	Parallel	Serial
Signals	Video data (NRZ) Ancillary data (blanking interval) Timing reference signals Identification signals	
Word length	10 bit + clock (8 bit + clock)	
Clock rate	27 Mbyte/s (8 bit) $C_B Y C_R Y$	216 Mbit/s (8 bit) 270 Mbit/s (10 bit)
Transmission	Bit-parallel/byte-serial	Bit-serial
Synchronisation	Timing reference signals in line blanking interval (with error protection)	
Interface	ECL push-pull signals D 25 subminiature	0.4–0.7 V_{pp}/75 Ω BNC (opt. interface)
Cables	9 twisted pairs alternatively: fibre-optic multiplex cables	Coaxial cables alternatively: mono-mode fibre-optic cable

Table 3.7 Signal parameters of CCIR Rec. 656

changed from 8 bits to 10 bits with compatibility being ensured. Coding of the serial signal data rate has been changed from the mapping code (243 Mbit/s) to the NRZI code (270 Mbit/s). Moreover, an extended interface format is being considered, which takes into account the enhanced requirements placed on postprocessing (e.g. higher chrominance resolution for colour correction).

3.5.2 Transition to digital studio techniques

Digital video signals in component form according to CCIR Rec. 601 with 4:2:2 studio standard yield a bit rate of 216 Mbit/s as a result of 8-bit coding for the luminance signal with a sampling frequency of 13.5 MHz and the two chrominance signals with a sampling frequency of 6.75 MHz. The distribution of this (net) bit rate in the studio complex calls for line connections with high transmission capacity and compact dimensions. For a radical change to digital studio techniques the studio needs a completely new installation. For a step by step transition, suitable studio areas (production complex) can first be digitised or individual digital equipment installed. For choosing the appropriate transition strategy, the arrangement of existing analogue studio areas plays an important part, whether CCVS coding is according to the PAL, NTSC or SECAM system or the analogue signals are component coded. The simplest case is a transition from analogue to digital component coding.

The studio bit rate of 216 Mbit/s net has to be carried on the connecting lines and in the studio equipment and distribution facilities, such as a crossbar matrix [160]. The main distribution matrix of the digital studio has also to be matched to a digital environment where signals are processed and transmitted at different bit rates. The bit rate has to be adapted to the 4:2:2 studio, usually for both directions of transmission. Table 3.7 gives two alternatives of the interface standard for the digital 4:2:2 studio. The EBU has specified an interface with bit-parallel or byte-serial transmission in its Document 3246-E 'EBU Parallel Interface for 625 line Digital Video Signals'. The 8-bit data words of the Y, U and V signals are transmitted time sequentially on eight parallel lines. This unidirectional connection of individual components in the 4:2:2 studio comprises nine twisted pairs of lines (18 wires) carrying the 8-bit data words and associated clock signal.

Serial transmission at 216 Mbit/s is bit-serial or byte-serial either via coaxial copper cable or via monomode fibre-optic cable.

3.5.3 Interface between TV studio and line network

The bit rates of the postal line network are defined in the form of a supranational digital hierarchy. TV signals at the fourth hierarchical level are transmitted at a rate of 140 Mbit/s and the third hierarchical level has

34 Mbit/s, while the introduction of a further level at 70 Mbit/s, is under discussion [159].

The reduction of the 4:2:2 studio data rate of 216 Mbit/s to the various hierarchical levels can be made according to two different criteria. One possible approach is to reduce the sampling rate, with only the active picture area being transmitted. A further approach is undersampling, although aliasing effects have to be tolerated. A third approach is by converting the sampling rate using appropriate prefiltering. In this case the resolution of the luminance and/or chrominance signals will be reduced.

The second basic method of bit rate reduction is by reducing the number of bits per sample using redundancy-reducing methods such as DPCM (differential pulse code modulation) or transform coding.

The reduction of the studio bit rate to 140 Mbit/s can be achieved without problems by restricting the transmission to the active picture area and by using 'light PCM' [155].

A special variant is the transition to the VBN (switched broadband network) of the German DBP-Telekom, which has been designed mainly for video conferencing. The VBN is a high performance digital fibre-optic network that was introduced in 1989. On the basis of the coding algorithm adopted by the CCIR, the Institut für Rundfunktechnik (IRT) has developed a 140 Mbit/s component codec which can be connected to the VBN via a user/network interface [156]. By restricting the studio signal to the active picture area, the data rate is reduced from 216 Mbit/s to 165.888 Mbit/s. The active picture area has 576 lines and 720 luminance or 2×360 chrominance samples per line. The data rate can be further reduced using an algorithm based on DPCM (differential pulse code modulation) while hybrid DPCM (HDPCM) is also used(Section 3.7.2).

With a data rate reduction to 140 Mbit/s, the picture signal is reduced to about 135 Mbit/s and the remaining transmission capacity reserved for sound, additional signals and for the pulse frame.

For the distribution of digital TV signals a transmission rate of about 70 Mbit/s is being considered, although this rate is not defined as a hierarchical level. The TV studio substandard of 108 Mbit/s would be appropriate for the 70 Mbit level. The required bit rate is achieved by noncoding of the blanking intervals using a DPCM algorithm. It remains to be seen whether the picture quality achieved and the postprocessing capabilities will satisfy the requirements of studio operation. The picture quality level at 70 Mbit/s will certainly be sufficient for the distribution to terrestrial transmitters via satellites or cable networks.

The transition to 34 Mbit/s for the sole purpose of signal distribution to transmitters and cable networks will also yield an acceptable picture quality. A reduction factor of this magnitude requires a sampling rate conversion for part of the studio bit rate with appropriate prefiltering of the signals. This is followed by the redundancy reduction to the desired bit rate of 34 Mbit/s.

3.5.4 4:2:2 studio equipment

The digital equipment of the 4:2:2 studio is taking on new dimensions.

Since the recording format for digital components was defined in 1986, the second generation equipment is now available. DCT (digital component technology) using 19 mm tapes is ready to be launched on to the market. While Ampex has opted for the 19 mm tape, 8-bit resolution and data rate reduction for DCT, Sony has gone ahead with the half-inch tape for Digital Betacam. The full-sized cassette has a capacity of 2 hours while the compact cassette, as used in camcorders, has a capacity of about 40 min. Sony uses a 'light' data rate reduction of about 2:1, the 10-bit technique according to the enhanced CCIR 601 standard yielding a total data rate of 240 Mbit/s.

Digital cameras were until now offered only by Panasonic and Ikegami, but the market leader Sony is soon to follow. Sony has developed a CCD (charge coupled device) element with 980 horizontally arranged active pixels which can be sampled at 18 MHz. Previously, Sony's 760-pixel CCDs used 14 MHz. The new 18 MHz read frequency matches the digital environment of future cameras. The 18 MHz corresponds to (4:3) × 13.5 MHz for digital component processing. In the analogue environment, the 18 MHz clock is also a good choice, since after appropriate coding with five times the colour subcarrier frequency of 3.58 MHz the signal correlates with the NTSC signals and after coding with four times the colour subcarrier frequency of 4.43 MHz with the PAL signals. The A/D conversion in Sony's future digital camera covers 10 bits or 12 bits. This digital camera will however also need some analogue components: at least two ICs for noise reduction and selectable video gain. Following the A/D conversion with 10 bits, further signal processing is fully digital. A convertor supplies a signal conforming with CCIR 601 [161].

3.5.5 Digital HDTV studio

The components of a fully digital HDTV studios are available today from European manufacturers. In 1989 BTS (Broadcast Television Systems) changed their development projects within Phase II of EUREKA 95 from analogue to digital techniques [162]. The key technologies for HDTV studio equipment have now reached near perfection making the equipment ready for operation on the HDTV studio standard 1250/50/2:1 with 1920 pixels being obtained per active line.

The first products from the new digital equipment generation were a digital HDTV disc recorder and digital CCD camera. Late in 1992 they were followed by prototypes of the first GBR (gigabit recorder) and a digital HDTV mixer, completing the equipment required for an all-digital HDTV studio. For the 1992 Olympic Games in Barcelona, BTS had already installed the first fully digital HDTV studio called 'Bolero' (Fig. 3.33).

The most important equipment of the production hardware is the video

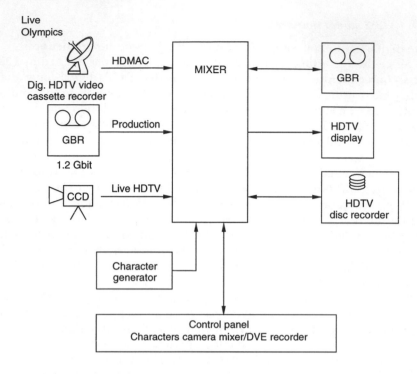

Fig. 3.33 BTS BOLERO HDTV Studio (BTS Olympic Editing Room)

recorder for 1.2 Gbit/s [162]. The gigabit recorder has the same operating functions as a good quality recorder for standard TV, e.g. visible fast forward and rewind search and slow motion. Approximately 1.2 Gbit/s have to be packed on the half-inch tape. Using the full 30 MHz bandwidth of the EUREKA 95 HDI (high definition interlaced) standard, 30 min of recording time is possible. Due to its data protection capability, the gigabit recorder is not only suitable for transparent HDTV recording, but also for general data storage or recording of HDTV signals with reduced source data. The benefit gained from the data reduction should be reflected in a longer recording time.

Parallel processing and recording is used to obtain the high recording data rate. For example, 2 × 8 heads are used each for recording and reproduction, plus two erase heads, totalling 34.

The gigabit recorder also operates according to the CDR (common data rate) method, the HDTV format developed in the US, which is a 1050/59.94 system. It corresponds to the data rate of EU 95, but the picture period is 16.6 ms compared to 20 ms as on the 50 Hz European system.

Another piece of digital studio equipment is the digital video mixer combined with a DVE (digital video effect) unit.

A digital field synchroniser is used in live recording or satellite broadcasting to synchronise both the mobile HDTV with the studio equipment. Although this instrument has analogue input and output signals, it is fully digital, with the luminance signal being sampled at 72 MHz and the colour-difference signals at 36 MHz, conforming to the EU 95 standard.

The example of the Bolero studio shows that the HDTV studio production can be fully digitised. It is an impressive demonstration of the equipment technology now available for future digital HDTV productions and systems and shows the high performance level of the European HDTV production standard.

3.6 Digital TV receivers

For the advance towards an all-inclusive digital TV technology from the TV studio via the programme distribution lines through to the media for feeding home receivers, it is not necessary to develop an all new technology solely for fully digital TV receivers. There is already a good platform provided by the digital TV receiver concept developed by ITT within their DIGIT 2000 system in the mid-1980s and DIGIT 3000 early in the 1990s, tried and tested with various receiver concepts in the industry.

The main motivation for a digital design of the TV home receiver was originally through better and more cost-effective manufacturing. Another advantage hardly considered at the time was that in future digital transmission systems there is no need to convert from and to the analogue video level, e.g. with the CCVS or MAC signal. After RF demodulation, the transmitted picture-coded TV signal can be directly applied to the picture decoder and screen display unit in the form of a serial datastream of, for instance, 4 to 40 Mbit/s depending on the picture quality.

3.6.1 Configuration of TV receiver

The main function modules of a standard TV receiver are the tuner and IF unit, video and audio processor and the deflection unit for beam current control in the CRT (cathode ray tube). The signal flow starts from the antenna and continues through the tuner RF section, the IF amplifier and the demodulator. These processing stages provide channel selection, amplification, filtering and demodulation of the channel signal. At the stage where the signal paths for the picture, sound and deflection signals separate, unmodulated baseband signals from other sources (e.g. from the video recorder or a digital video disc) can be fed to the home receiver. The basis for digitisation of the TV receiver lies in the video, audio and deflection sections. The RF and IF unit is currently of analogue design, as are the amplifiers for the RGB picture signal, the stereo signal and the high-voltage unit for vertical and horizontal deflection (Fig. 3.34).

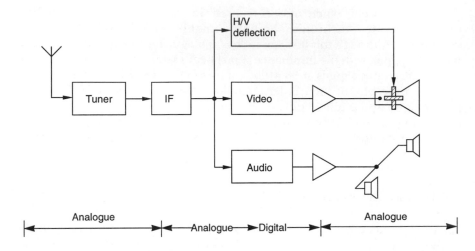

Fig. 3.34 Transition to digital technique in TV receiver

3.6.2 Picture memory

The following known system inherent disadvantages of the established TV systems PAL, SECAM and NTSC are:

- visible line structure of the picture, of PAL with 625 lines and especially NTSC with 525 picture lines
- large-area flicker due to low frame frequency of 25 Hz (field frequency 50 Hz) with PAL and SECAM
- edge flicker due to 2:1 deflection of field frequency
- cross-colour and cross-luminance due to imperfect separation of the luminance and chrominance signals transmitted in frequency multiplex
- poor sound quality due to intercarrier reception (Section 3.3.1), particularly apparent in stereo broadcasting according to the dual-carrier method. As the SECAM system sound is amplitude modulated, stereo broadcasting has not yet been introduced.

Regarding the picture area, the picture memory plays an important part in reducing or solving these problems. Interestingly the following additional functions can be implemented:

- digital comb filters avoid cross-colour and cross-luminance effects
- improved contour sharpness, no visible line structure and edge flicker
- no large-area flicker caused by picture memory readout and display of the TV picture with 100 Hz vertical frequency
- noise reduction by field linkage
- new features such as still picture, zoom, slow motion, picture-in-picture or picture graphics from teletext or user interface [1].

The functions mentioned above are limited to the receiver end using introduced transmission methods. A picture quality improving effect of the picture memory is sometimes also referred to as IDTV or ADTV improved/advanced definition television). Depending on complexity, the picture memory requires a capacity of 2.5 to 25 Mbit. For the realtime transfer of the picture information to and from the picture memory, clock frequencies of over 50 MHz have to be used. In view of the costs involved, the picture memory is only available in the high-end receivers. For economic reasons it is preferable to adopt a strategy where the quality improving effects of the picture memory are already contained in the digitally transmitted signal.

3.6.3 DIGIT 2000 system

As early as 1983 the first ICs of a TV receiver DSP chip system were introduced by ITT Intermetall. Today, the chip set DIGIT 2000 is fully developed and mainly used in high-performance TV receivers.

The main features of the DIGIT 2000 system are the decoding and processing of all TV picture and sound transmission standards, processing of data services such as teletext, picture correction by special algorithms and filtering methods and the reception of scrambled satellite programmes. The IC set is also compatible with old and new technologies such as camcorders, video recorders, CDs and video discs.

The VLSI circuits combined under the term DIGIT 2000 together with a small number of conventional components reduces the component count that was previously required for the control, video, deflection and sound switching circuits. The benefits are: fewer external components, no tolerances, no drift and ageing, programmability, operating convenience supported by software, processor-aided adjustment of the receiver in manufacture, TV sound for all introduced standards and being multistandard for PAL, NTSC and SECAM.

Fig. 3.35 shows the block diagram of a TV receiver based on ITT-DIGIVISION. The picture and sound signals with the operational and control functions are handled digitally. After conversion back to the analogue level, the TV signals are routed via power output stages to the CRT and to the stereo loudspeakers. The data transfer between the ICs is handled by a 7 bit bus sampling at 17.7 MHz, hence the 123.9 Mbit/s data rate on this bus. The clock generator, operating at four times the colour subcarrier frequency (17.7 MHz), provides the signal processors with the clock signals $\Phi1$ and $\Phi2$ of opposite phase.

The DIGIT 2000 system comprises the following ICs [164]:

The CCU (central control unit) contains an 8-bit microcomputer and co-ordinates all processes within the digital system. It stores the factory preset tuning values to ensure optimum picture reproduction. The CCU also

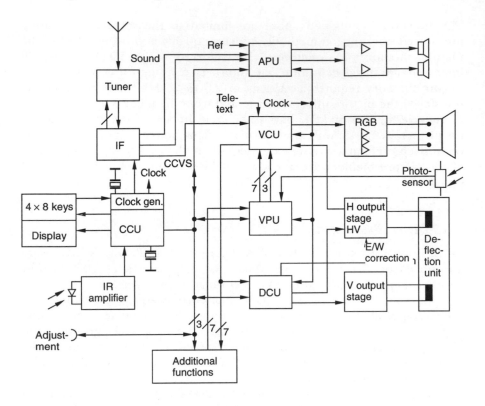

Fig. 3.35 TV receiver based on system DIGIT 2000 (source: ITT)

processes the infrared commands received from the remote control unit such as viewer preferences for channel selection, volume, brightness, contrast and colour saturation, before passing on to the other functional stages. The CCU has the following main functions:

- infrared remote control
- 32 commands for direct control
- PLL (phase locked loop) frequency synthesiser tuning
- nonvolatile programme storage
- LED channel display, DSP control for video, sound and deflection via serial IM-bus.

The chip is designed in *N*-channel MOS technology and contains an 8-bit microcomputer and an EAROM in floating gate technology with a capacity of 96 words of 8 bits each.

The VPU (video processing unit) contains the functions which within a conventional colour TV set are performed by the colour decoder (e.g. PAL decoder): retrieving the colour difference signals, colour killer, phase comparison for colour subcarrier and PAL delay line. The VPU is a real-time digital signal processor designed as an *N*-channel MOS circuit. The chip comprises the following functions: code convertor, chrominance filter, colour subcarrier trap with focus control, contrast multiplier with limiter for the luminance signal and colour signal processing with, for example, automatic gain control, colour killer or PAL identification.

The VCU (video codec unit) operates in conjunction with the video processing unit. The analogue/digital with the digital/analogue conversion of the colour picture signals is performed by this chip. Moreover, this chip contains the RGB matrix (red, green, blue) with conversion of the YUV representation into RGB signals. It also has inputs for videotext and teletext and programmable auxiliary circuits for blanking, brightness control and CRT adjustments.

The VCU is a bipolar IC in 3D technology. The following functions are programmed into the silicon chip: video amplifier, A/D flash convertor for the CCVS, D/A convertor for the Y and R-Y, B-Y signals and the RGB matrix.

The audio ADC (analogue digital convertor) converts the analogue input signals into digital signals in two operations. First, the audio signal is converted by a modulator into a pulse code modulated signal. In the subsequent digital sound filter, the resulting 1-bit datastream with a maximum rate of 4 MHz is reshaped for a word rate of 35 kHz per channel. The data word length is 16 bits. Due to the high sampling rate of the pulse code modulators, an anti-aliasing lowpass filter is not required at the input. This kind of A/D conversion causes an S/N ratio corresponding to that of a 13-bit A/D convertor.

In the APU (audio processing unit) the sound signals are processed totally digitally. The APU is a programmable real-time DSP for the digitised sound signal furnished by the ADC. APU and ADC form a complete stereo sound processing system. Although the basic functions are mask programmed, they can also be programmed via the serial bus input. This allows the system to be used for other sound TV standards in addition to the dual-sound carrier method. The APU is designed as an *N*-channel MOS circuit and performs the following software controlled functions: dematrixing and de-emphasis, physiological volume control, pseudo stereo, decoding of mono, stereo and dual-language mode identification.

The DPU (deflection processing unit) extensively controls the vertical and horizontal deflection of the electron beams in the CRT. The chip, a programmable VLSI circuit designed in an *N*-channel MOS technology, performs the following deflection functions: video clamping, sync. separator, H and V synchronisation, and sawtooth generation with east-west correction.

In the clock generator IC the clock signal for the digital TV receiver is derived in an integrated bipolar circuit. The clock signal is a nonoverlapping two-phase clock with four times the colour subcarrier frequency, i.e. 17.73 MHz for PAL and 14.32 MHz for NTSC.

3.6.4 DIGIT 3000

The DIGIT 2000 concept, a self contained TV-oriented IC system, has been succeeded by DIGIT 3000, which is designed as an open-architecture digital system for universal applications. DIGIT 3000 is therefore suitable for use in commercial and industrial new network systems. The open architecture of the DIGIT 3000 is implemented by building blocks integrated in a system IC. The silicon blocks are assigned clearly defined functions, e.g. microcontroller, multistandard video processor, text and graphics processor or A/D–D/A convertor [163]. The IC design can be varied via the building blocks and the functions quickly and economically adapted to specific customer needs and applications. There are no expensive redundancies in the ICs. Table 3.8 gives an overview of the video and audio building blocks. Further building blocks are intended for text, graphics, data, control and special functions. The building blocks — concentrated on one or several chips — can be universally used by the consumer electronics industry for specific circuits.

Video	Audio
8-bit A/D 10-bit D/A	16-bit A/D 16-bit D/A
Multistandard decoding	User-programmable DSP
Display processing	IF selection/demodulation
Satellite TV (MAC)	Adaptive de-emphasis (Wegner-Panda)
Interfaces with digital systems (DVI, CD-I, PC)	Multistandard baseband processing
Multistandard coding	Interfaces with digital systems (DAT, DCC, CD)
Picture storage (for format conversion, filtering)	Features (stereo sound, Radio Data System RDS)
Features (picture in picture, zoom, etc)	Data reduction (for DAB, modem applications)
PICTUREbus interface	SOUNDbus interface

Table 3.8 Extract from ITT library of building blocks for application-specific chips for implementing DSP in consumer products (further blocks available for text, graphics, data, control and special functions)

The concept of the building block can be expanded to the chip level since it uses only one control clock and versatile control and data bus structures. The building blocks of the DIGIT 3000 IC system communicate via the following bus systems:

- picture bus: a digital orthogonal (16+3)-bit bus for picture signals. It can be configured for instance as YUV or RGB of variable bit lengths.
- sound bus: a digital, serial 3-wire bus for sound signals which is able to transmit up to four channels between the audio processors.
- DIGIT bus: a serial 1-wire bus with variable protocol structure for system control by the system controller.

Thanks to the standard submicron CMOS technology, this software-supported open-architecture system is suitable for the introduced video, audio and computer standards especially for combining analogue and digital circuits on one chip (mixed mode design).

All circuits are designed in 0.8 µm layout for mass production. The chip integration of analogue and digital circuit components brings further advantages for the various applications: one supply voltage only, short connecting lines avoiding interference, and chip integration of processors, convertors and sensors.

3.6.5 Future generations of TV receivers

3.6.5.1 Multimedia

Generally, multimedia systems are understood as the components of computer technology, telecommunications and consumer electronics. Specifically, the PC or TV set has to be put into the centre of a system that processes video and audio signals or transfers computer data. Depending on the applications and signal processing requirements implemented in the multimedia system, different system configurations are obtained. For instance, the video data rate can be converted via an adaptor, so that the different formats, e.g. PAL signals and VGA (video graphics adaptor) graphics can be displayed and superimposed on a computer monitor.

A multimedia system may also contain an optical mass memory such as CD-ROM or CD-Interactive. Using data companding (compression and expansion), motion pictures can be manipulated on the computer in addition to still pictures. An interface allows connection to Telecom data networks so that the system can be used as a terminal for video conferencing.

3.6.5.2 HD-MAC receivers

Provided the relevant standard will be implemented, the future HD-MAC receiver must be able to expand in real-time the compressed picture signals received and to reconstruct the TV picture with the aid of motion vectors. In addition, it must be able to process the data packets and data information for sound, text or control. The receiver should be downward compatible and able to integrate new signal sources. The HD-MAC receiver must satisfy the multistandard requirement for HD-MAC, D2-MAC and NTSC, SECAM and PAL and additionally for the different horizontal and vertical frequencies. The degree of complexity of the HD-MAC receiver is determined by the clock frequencies according to the requirement for multistandard compatibility: PAL, SECAM, NTSC: 13.5 MHz; D2-MAC 20.25 MHz; HD-MAC 77 MHz [1].

3.6.5.3 Picture-processing receivers
The use of digital VLSI technology in the receiver does not only present a multistandard TV set for NTSC, PAL and SECAM, but also allows picture processing with higher resolution and greater aspect ratio (16:9). According to US forecasts, by the year 1999 more than 50% of TV receivers will be equipped with picture memories, DSPs, microcontrollers and DRAMs.

Based on psycho-optical effects, these components may cause a virtual quality improvement without the need for changing the infrastructure of the terrestrial transmitter networks (satellite, CATV) to digital HDTV. In the US it is widely believed that this type of TV receiver will pre-empt HDTV receivers with a corresponding digital network structure [165].

3.6.5.4 DSP development for HDTV receivers
The future high definition video processing in TV home receivers calls for a change from the single processor system (Harvard architecture) to parallel processor configurations. With further development it will be possible to implement processor configurations with more than 10 GOPS (giga operations per second) as are required for freely programmable real-time video with clock frequencies of >100 MHz. This means a capacity increase of 10^3 to 10^4 compared with present-day microprocessors. The design of such a parallel or array processor has already been presented under the project name Data Wave [1]. This processor is made up of 16 cells linked by 12-bit buses. The individual cells contain the multiplier/ALU (arithmetic logic unit) and shift register and programme memory. The concept is based on a 0.8 µm layout and it is intended to halve its current size so that 64 cells can be accommodated on the same area. With the resulting processing power of 32 GOPS at for instance 250 MHz clock frequency a freely programmable HDTV-compatible video processor could be used.

Further technical development of digital TV signal processing will lead to the single-chip system being introduced through economic pressures and because these high data rates are limited by the chip's physical boundaries. Using a programmable open processor architecture, the hardware determines the basic functions of consumer products while the user selects the features through the system's software.

DSP will be extended to include the RF and power stages and VLSI DSPs will also be used in IF processing and tuning. Suitable algorithms for digital demodulation methods have already been devised with development similar to that used in computer technology [1].

3.7 Video coding

3.7.1 Psycho-optics

The following description of some of the phenomena in psycho-optics shows that the inherent weaknesses of the human eye can be utilised for picture coding by data compression.

3.7.1.1 Optical illusions

When viewing a part of a picture, the human eye also takes account of the surrounding area which makes the eye sensitive to grey level changes in the neighbouring areas and with medium frequencies being preferred to the high or low frequencies. The decreasing sensitivity of the eye at high spatial frequencies has a twofold advantage during picture coding: firstly, picture contents corresponding to higher spatial frequencies are subjected to coarse quantisation, and secondly, coding errors within the limit frequency range of the eye's resolution will not be perceived. This filter function increases the contrast to make sharper edges [166].

The following illustrations give some examples of optical illusions:

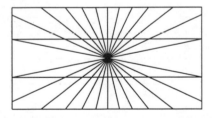

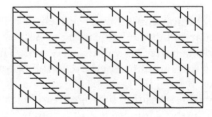

Fig. 3.36 Optical illusions

The eye can also be deceived by spatial representation. The ball positioned further back in Fig. 3.37 appears to be distinctly larger than the one in the foreground, although they are of equal size. The impression of spatial depth is created by the converging lines.

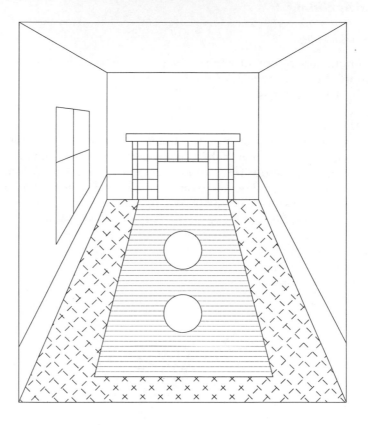

Fig. 3.37 Optical illusion caused by spatial impression (ball; the carpet in the foreground appears to be narrower than the wall in the background) [166]

Another optical illusion caused by grey areas is referred to as the Mach effect. At the boundaries between adjoining areas the impression of a shift towards lighter or darker shading values is created. This effect can be seen at best when looking with an unfocused eye at the boundary between a light and a dark area (Fig. 3.38) [170].

It appears as if in the transitional range from black to white there would be a brighter stripe in the white area and a somewhat darker stripe in the black area. This effect is used for video coding by adopting a coarse quantisation for the brightness transitions which go unnoticed by the eye. This is particularly important since the actual information in the pictures is

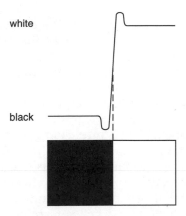

white

black

Fig. 3.38 Test pattern for Mach effect and perceived brightness

not contained within the areas but in the border and transitions. By accentuating the object edges, a sharper picture can be acheived without registering any impairment.

3.7.1.2 Pattern recognition

The working of the human brain has a considerable influence on a viewed picture. Incomplete object structures are perceived and automatically compensated for to give the impression of a complete picture (Fig. 3.39). Another characteristic feature of the eye is pattern recognition: the brain ignores insignificant details provided that the essential picture information is present. Distortions will not be noticed if they have little influence on the pattern recognition and will only be noticed on closer study of the still picture.

Fig. 3.39 Recognition of patterns (white triangle) [166]

3.7.1.3 Oblique effect

The human optical system adapts itself to the environment by focusing on horizontal and vertical shapes and not very sensitive to others. This so-called oblique effect (Fig. 3.40) is used to advantage in picture coding: picture signal components with diagonal spatial frequencies can be subjected to a coarser quantisation. It should, however, be noted that the eye becomes highly sensitive for slight deviations in a specific pattern.

In addition to the spatial characteristics of the eye, the temporal characteristics are exploited for picture coding. If, for instance, a cutout is shown from a full-motion scene, the eye will not be able to recognise any details. In terms of picture coding this means that moving objects can be quantised with a lower spatial resolution. This effect becomes less important when the standard 4:3 PAL system changes to the HDTV system with 16:9 aspect ratio. An example of this is when you are in a room viewing through the window the street outside away from the window. If you move close to the window and look into the street, the angle of view broadens to roughly a 16:9 aspect ratio. In other words, peephole viewing is turned into telepresence.

Fig. 3.40 Recognition of (errors in) patterns, oblique effect (here: displacement and distortion) [171]

3.7.2 Video coder

Data compression is attained through video coding. Digitised pictures supplied by the TV camera are subjected to data compression. These pictures are sampled in pixels and the luminance signal Y and the chrominance signals R-Y and B-Y of each pixel are each assigned a value (Fig. 3.41). This PCM coding uses purely linear binary numbers initially. For the resulting bitstream, a data rate of 160 Mbit/s is obtained for the PAL signal with 720 pixels per line and 576 active lines at a frame frequency of 25 Hz. With HDTV, the vertical and the horizontal resolution are doubled and a new aspect ratio of 16:9 is used compared to the previous 4:3. For HDTV, the sampling frequency of the luminance signal will be increased by a factor of $2 \times 2 \times (16:9)/(4:3)$ to 72 MHz as compared to 13.5 MHz stipulated in CCIR Recommendation 601. A total data rate of 1.152 Gbit/s is obtained for the video source signal in Y-U-R representation [168].

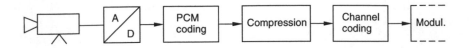

Fig. 3.41 Functional modules of video coder

The source bitstream has to be drastically reduced by data compression. Since the data signal taken to the receiver via a transmission medium is subjected to interference (terrestrial propagation, coaxial cable, satellite), a certain bit capacity has to be added to the datastream for error protection. Moreover, the transmission media features different bandwidths: for terrestrial propagation, the VHF channel bandwidth is 7 MHz, the UHF bandwidth 8 MHz, and with CATV it is 7 MHz in band III and 12 MHz in the hyberband. The satellite transponder has a minimum bandwidth of 27 MHz. Since terrestrial distribution is the most critical case — in the USA the bandwidth of the TV channels is 6 MHz only — the following conditions have to be met:

- The compression factors for HDTV lie in the range between 30 and 60, while the corresponding factors for standard TV are determined by the required quality.
- Regarding broadcasts in which viewers can participate, the coding and decoding algorithm should not cause any noticeable temporal delays [169, 170].

3.7.2.1 Redundancy and irrelevancy reduction
For choosing a suitable method of compression, the characteristics of both the picture source and the human eye play an important part. For redundancy reduction, temporal, spatial and statistical relationships between the picture elements are used:

- In a sequence of pictures the consecutive pictures are similar with the changes not the result of new content but mainly of the spatial movement of patterns (Fig. 3.42).
- Natural pictures have a large amount of areas and edges. Therefore, neighbouring picture elements (pixels) are in all probability similar.
- With a camera operating at optimum performance midway grey levels occur far more frequently than the extreme white and black levels.

The above effects describe the similarity of the pixels within a specific picture and of consecutive pictures. It is this spatial and temporal redundancy that has to be used in data compression.

A simple and basic approach is to transmit not the complete information, but only the changes. Superfluous information will be omitted in the datastream, causing redundancy reduction. The receiver constructs a fresh picture with the original picture content and incorporating the changes.

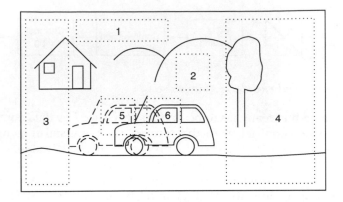

Fig. 3.42 Picture contents   1,2   *Spatial redundancy*
                                    1,2,3,4   *Temporal redundancy*
                                    5,6   *Temporal prediction*

Another approach to save bits is to transmit differential pictures. With such pictures, low levels occur more frequently than high levels and therefore can be transmitted with short code words, whereas longer code words are used for larger differences. For reasons of frequency distribution, data transmission capacity can be saved (Fig. 3.43) [169].

The purpose of redundancy reduction in the video coder is to confine the information content of the source — the entropy. In communication theory, entropy is a vital factor according to the Shannon theorem: the average codeword length of a source cannot be smaller than the entropy of the source [187]. The approximation of the average data rate to the entropy is therefore a measure of the code efficiency or a reciprocal measure of the code redundancy. Due to its distribution density, a differential picture has for instance an entropy of about 3 bits at small amplitudes. With increasing brightness, the picture has an entropy of about 8 bits, since on average all grey levels occur equally frequently [172].

Irrelevancy reduction can be understood in this context as the toleration of nonvisible coding errors, using the limits of optical perception. For irrelevancy reduction, high spatial frequencies are suppressed with a coarse quantisation applied but this is an irreversible process with losses. The decoded picture differs from the original one, but the loss in quality is subjectively not perceived. Irrelevant information can also be eliminated in the frequency domain, relying on the fact that on the average, the pictures have less high frequency components and the eye is less sensitive at high frequencies. For a picture line with constant brightness a frequency of 0 Hz is only transmitted.

The transformation to the frequency domain is originally a reversible

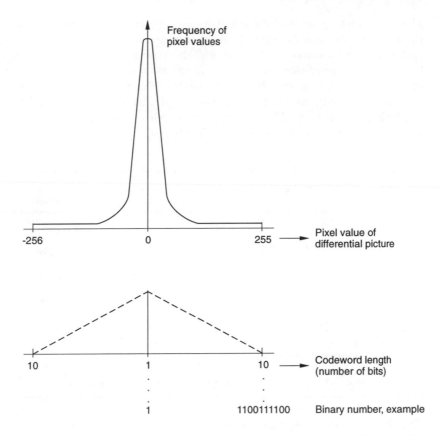

Fig. 3.43 Utilising the frequency distribution of a differential picture for data compression [169]

operation and free of losses. Although the suppression of the high frequency components of the signal causes an objective loss in quality, this loss is not perceived because of the irrelevancy reduction. From the above assumptions it can be deduced that the bit rate reduction may or may not cause a noticeable loss in quality. With a view to subjective perception, the coding errors should be concealed in the picture, and if the coding errors are invisible one can truly speak of an irrelevancy reduction.

The levels from a differential coder or a transform coder have to be quantised, the signal amplitudes being assigned decision thresholds. The resulting substitution level will be coded with a fixed or variable word length with the reciprocal process performed within the receiver and the resulting error called a quantisation error [173].

The quantised coefficients from a differential PCM or a frequency transformation have a high probability of having zero value for many points.

Moreover, large groups of these zero values are stringed together. Coding of these zero runs can be simplified by stating the zero length (run-length coding). The nonzero coefficients and the zero run in a line may be used for further compression, since certain values occur more frequently than others. The bit rate can be significantly reduced by allocating a short codeword to frequently occurring sequences and long codewords to rarely occurring sequences. The most popular video coding method is the Huffmann code introduced in 1952 (Table 3.9).

Symbols	Probability	Huffmann code	Code word	Code word length
a	2^{-1}	00	0	1 bit
b	2^{-2}	01	10	2 bits
c	2^{-3}	10	110	3 bits
d	2^{-3}	11	111	3 bits

Example: fixed code rate: 2 bits/sample
         average Huffmann code rate:

$$= 2^{-1}(1 \text{ bit}) + 2^{-2} \cdot 2 + 2^{-3} \cdot 3 + 2^{-3} \cdot 3 = 1.75 \text{ bits/sample}$$

Table 3.9 Example of Huffmann coding [195]

3.7.2.2 Coder peripherals

The heart of source coding consists of two parts, the preprocessor and the coder. Initially, the HDTV studio signal is preprocessed so that by a reduction of the sampling frequency from, for example, 72 MHz in the studio to 54 MHz for the transmission link a first data rate reduction can be made. A further reduction can be achieved by reducing the chrominance resolution. By taking advantage of the human psycho-physical perception limits in the field of chrominance resolution, the U and V chrominance components of the video component signal can, for instance, be reduced by a factor of four in the horizontal direction and by a factor of two in the vertical direction. By using this method in the preprocessor, the data rate for HDTV can be reduced to approximately 500 Mbit/s without any significant subjective loss in quality [194, 167].

Every video coder needs a data memory at the output for transforming the variable data rate, from that of the Huffmann coder, to a fixed data rate to one required for the channel transmission. The data buffer is designed as a FIFO (first in/first out) memory. The memory capacity must take account of the variability of the data rate of a complete picture. To avoid overflow of the FIFO buffer, a reverse adjustment in the coding process has to be made. This is usually done in the quantisation phase. In a special case, the so-called filler bits may be inserted to avoid the data rate from dropping below a predefined threshold [194].

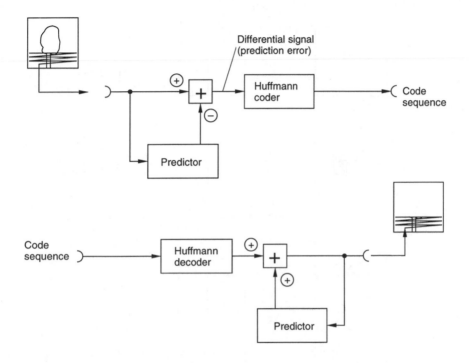

Fig. 3.44 Coder and decoder for loss-free DPCM [174]

3.7.2.3 Differential PCM

For coding it is first necessary to limit the frequencies in the picture sequence through a filter. This process yields an amplitude distribution that is closely concentrated about zero. Bit-saving entropy coding is used for transmission. The decorrelation filter derives the output signal from the difference between the input signal and the prediction signal (Fig. 3.44). This method has the advantage that the predictor can be realised simply. From the pixel transmitted, the predictor must derive the prediction for the next pixel to be transmitted. The better the match between the real input signal and the prediction, the lower is the bit rate after differential coding. The differential signal is frequently referred to as prediction error, since it

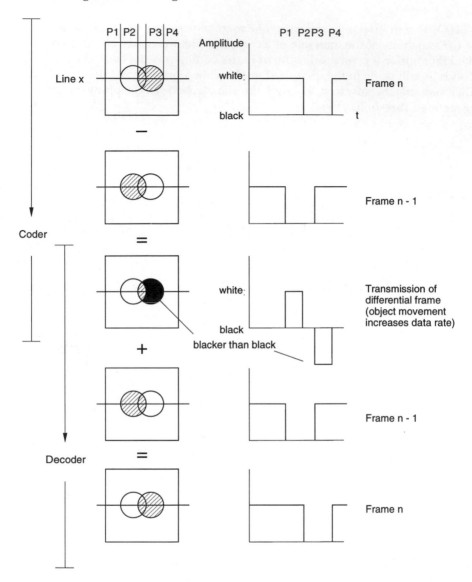

Fig. 3.45 Interframe DPCM without motion compensation

corresponds to the difference between the prediction and the pixel transmitted. In conjunction with the receiver, which also derives its prediction from the received pixels and corrects it by the transmitted prediction error, a loss-free DPCM is obtained (Fig. 3.45). Since the loss-free DPCM with an entropy of approximately 3 to 4 bits yields a compression factor of about 2 only, a lossy DPCM has to be used. For this purpose, the prediction error has to be quantised; with the resulting quantisation error should be negligible for redundancy reduction [174, 176].

3.7.2.4 Prediction with motion compensation

The TV systems in use are based on line interlacing, that is two fields are transmitted at 50 Hz to obtain a frame frequency of 25 Hz. Therefore, distinction is made between intrafield predictors operating within one field and intraframe predictors deriving the prediction from a frame that is composed of two fields by a frame memory. The prediction is improved by reference to previous frames in addition to the current frame. One speaks of interfield prediction if fields of the same parity are always used whereas for interframe prediction any (even or odd) fields may be used [174].

With a pure prediction method, the background shown after an object has moved cannot, of course, be predicted. To close this information gap, a higher data rate is briefly resorted to. A better solution is the combination of predictive and interpolative coding. Interpolation frames are obtained from intraframes and prediction frames, i.e. the sequence is intraframe, interpolation frame, prediction frame, etc. [167].

The real breakthrough to coding methods with high compression factors came with motion-compensated coding (Fig. 3.46). In the first step of data reduction, only those pixels of a new picture are transmitted according to the DPCM method with prediction which have undergone a change due to the movement of the object. The unchanged pixels are redundant and can be reconstructed by the frame memory in the receiver. For a further reduction of the pixels to be transmitted, a displacement vector is used to indicate the block displacement of a moving object from frame to frame. Through the displacement vector for a pixel, the receiver can retrieve the block displacement vector for the moving object from the memory. The information on the freed background has of course to be transmitted too. This method is referred to as interframe DPCM with motion compensation [191].

To estimate the displacement vector on a frame-by-frame basis, the block matching method is usually employed. For the application of this method in video coding, chip solutions are already available. For each block consisting for instance of 16 × 16 pixels, the block with the best congruence is searched for in the previous frame. To limit the search algorithm, matching is not performed in the entire previous frame, but only within a certain search area. The larger the search area in the horizontal and vertical directions, the more complex becomes the estimation of the displacement vector (Fig. 3.47) [167]. An ideal motion compensation would yield differential frames with practically no information (all-black picture). For a detailed description of the movement an appropriate amount of data would of course be required. It does not yet allow precise identification and measurement of all types of object movements.

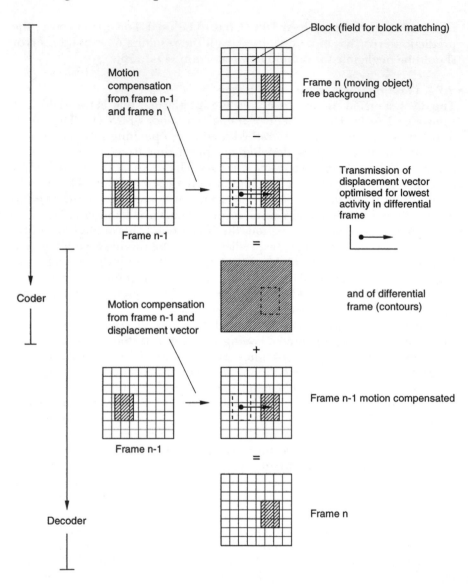

Fig. 3.46 Motion compensation with block matching

Despite imperfect motion compensation, the data rate for the differential frame achieved with the block matching method is much lower than with pure DPCM with prediction. DFD (displaced frame difference) is applied to determine the degree of agreement between the moving frame and the corresponding previous frame. The absolute difference values from the current block and the previous block are summed up. The smaller the results, the better the DFD. Interframe DPCM with motion compensation allows a data rate of 0.5 to 2.8 bit/pixel to be achieved for the frames, depending on the picture to be transmitted [175]. Since this data quantity is too high for terrestrial broadcasting of digital HDTV, a further data reduction measure is required and is implemented by a transform coding, or more accurately a combination of interframe DPCM with motion compensation and transform coding.

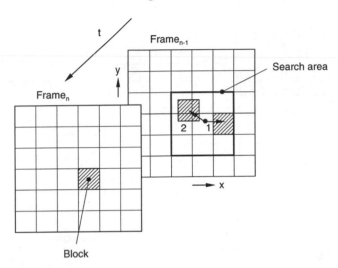

Fig. 3.47 Prediction with motion compensation: result of estimation is the displacement vector (e.g. 1 or 2) with x, y co-ordinates and accuracy of 1 or $\frac{1}{2}$ pixel

3.7.2.5 Discrete cosine transform
The most popular decomposition technique in the frequency domain is DCT (discrete cosine transform), which is a derivative of the discrete Fourier transform. With DCT, a block of the differential frame with motion compensation, consisting typically of 8×8 or 16×16 pixels, is transformed into the corresponding frequency domain. To simplify the calculations, fast algorithms are used similar to Fourier transform [187]. A special feature of DCT is that stochastic signals with autocorrelation functions are described by approximately linearly independent spectral values. The resulting coefficients are quantised and coded for transmission. With this additional transform coding, the masking effect of the human eye is used and high-frequency coefficients are subjected to a coarse quantisation or assumed to be zero. This is possible because in the neighbouring pixels with similar values few of the coefficients will have significant values at low frequencies. The coefficients at higher frequencies have a value of zero or close to zero to be stringed in a zero run (Fig. 3.48) [189, 190].

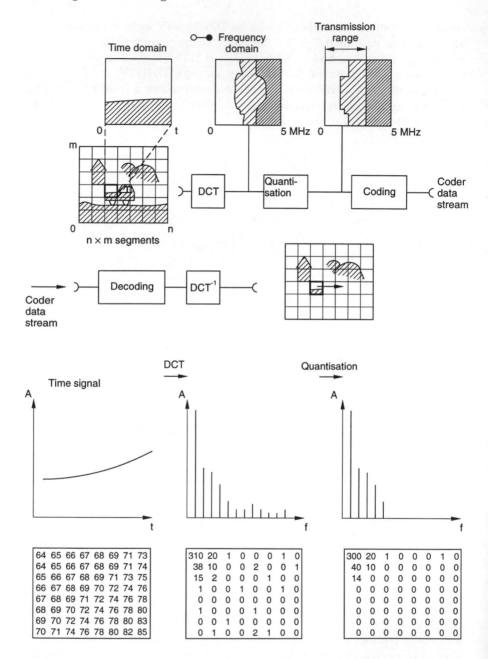

Fig. 3.48 Data compression by DCT [190]

Transform coding uses the weakness of the eye, of being less sensitive to high frequency picture components. Therefore a coarser quantisation or quantisation with zero can be used for the coefficients at higher frequencies.

In high-activity picture areas it is difficult to detect coding errors (irrelevancy) and it should be considered that the two effects described may cancel each other out since in low-activity picture areas coding errors become very evident with a higher bit rate used for coding.

Note: General DCT uses a block of $N \times M$ samples from the frame as the input coefficient. The algorithm is similar to that of the discrete Fourier transform (Section 4.7.4), with the special feature that the DCT output coefficients are real. The block is transformed by the following representation [187]:

$$c(p,q) = c_o(p,q) \frac{2}{\sqrt{MN}} \sum_{n=0}^{N-1} \sum_{m=0}^{M-1} s(n,m) \cos\left[p(m+\frac{1}{2})\frac{\pi}{M}\right] \cos\left[q(n+\frac{1}{2})\frac{\pi}{N}\right]$$

$$\text{with } c_o(p,q) = \frac{1}{\sqrt{2}} \text{ for } p \neq q, p \times q = o$$

$$\frac{1}{2} \text{ for } p = q = o$$

Subband transformation is an alternative to DCT. This method is in some ways even better since it is not limited to one block and avoids the errors known from DCT, such as block structures [167].

3.7.2.6 Hybrid DCT

Hybrid, motion-compensated DCT coding combines two powerful compression methods to give the required amount of data reduction. The first method, which is based on motion-compensated DPCM, uses temporal correlation, while the second method based on discrete cosine transform uses spatial correlation. The special feature of hybrid DCT is that the differential frame is subjected to transform coding. Fig. 3.49 shows the principle of hybrid DCT. Since the coder needs the quantised DPCM frame in the time domain, in the feedback loop an inverse DCT is obtained before the summing point. The same holds true for the decoder (Fig. 3.50).

Hybrid DCT also has some disadvantages. Occasionally it may be necessary to decode consecutive pictures independently, such as in picture editing. Another problem is fast search running with digital VTR (video tape recording). A fast flyback causes problems with hybrid-coded pictures, since the current and the previous picture are interchanged.

For these reasons, intraframe coding is preferred to hybrid DCT in picture postprocessing. To combine the advantages of hybrid DCT and intraframe DCT, whole pictures can be intraframe-coded at fixed intervals. The ISO-MPEG standard is based on this concept (Section 3.7.2.10) [177, 178, 179].

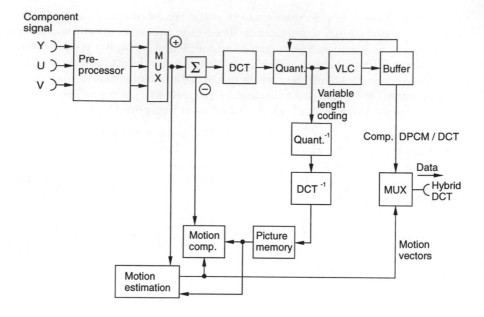

Fig. 3.49 Block diagram of coder for hybrid DCT

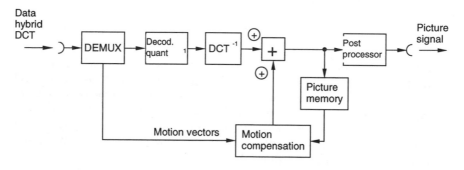

Fig. 3.50 Hybrid DCT decoder

3.7.2.7 Vector quantisation

Vector quantisation is another coding method still in development. Unlike previous quantisation methods which refer to individual pixels and are therefore designated as scalar quantisers, vector quantisation refers to a group of pixels which are combined in the input vector (Fig. 3.51). The input vector is a group of neighbouring pixels. An equivalent vector for the input vector is looked up in a codebook. The equivalent vector is the vector in the code book coming closest to the input vector. This code vector is assigned an appropriate code which is transmitted. In the decoder, the code word received and the built-in code book are used to find the equivalent vector and to insert it in the appropriate place in the picture. The code book is a set of elements allowing reconstruction of the given picture sequences. The bounds of vector quantisation are still determined by the available memory capacity and computational power. It can be assumed, however, that vector quantisation will gain more importance in the future [181].

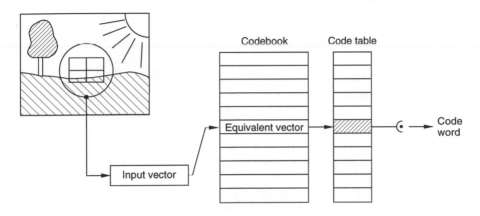

Fig. 3.51 Principle of vector quantisation [181]

3.7.2.8 Subband coding

Subband coding is related to transform coding and is used in audio coding according to the MUSICAM method (Section 2.4.2) where psycho-acoustic phenomena such as the masking effects and the resting threshold are optimally employed by subband coding. The masking effects of the eye in the frequency domain cannot be used for picture coding. Only the poorer spatial sensitivity of the eye at higher frequencies with the redundancies of the picture signal in the visual system can be used advantageously by subband coding. Another advantage is the relatively simple generation of a low-resolution picture due to the low-frequency subbands. This hierarchical coding concept can be employed in HDTV broadcasting. The lowpass content of the HDTV picture is transmitted as an EDTV (enhanced definition TV) picture (Fig. 3.52). The quantisation error occurring in EDTV picture coding in the lowpass content is determined, separately coded and transmitted together with the high frequency content of the HDTV picture while in the HDTV receiver both the EDTV signal and the residual signal is used for reconstructing the picture. In the case of a relatively simple receiver or of a disturbed transmission link it is possible to only display the EDTV picture [185].

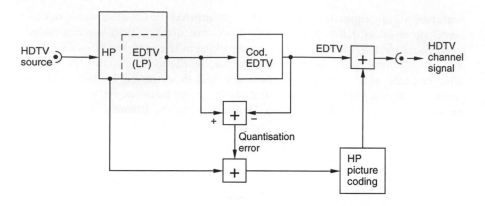

Fig. 3.52 Hierarchical EDTV/HDTV coding in conjunction with subband coding

For subband coding, the picture signal is split into its spectral components by bandpass filters. If there is no bandpass overlapping, it can be assumed that the output signals are uncorrelated, that is they do not exhibit any linear interdependencies. Every bandpass signal can be subjected to scalar quantisation and coded. With a small bandwidth of the subbands, the power density spectrum in the subband is flat so the samples are relatively independent of each other. Subband division is an advantage where the power in several subbands is low and thus only requires a coarse quantisation, if any. The power concentration in the low frequency subbands as in the case of live pictures is to be used *a priori* and as such does not prove to be an advantage for transform coding.

Without the use of a data reduction method, subband coding has no influence on the required bit rate. To ensure equal sampling accuracy, each subband has to be provided with a sampling rate corresponding to that of full-band coding divided by the number of subbands. When adding up the bit rates of the subbands, the same bit rate is obtained as for coding without the subband division. It is the decimation of the subbands that brings the bit rate reduction (Fig. 3.53).

The components of subband coding are: bandpass filters, decimation, scalar quantisation, coding and addition of subband bit rates. Subband coding was initially applied to the intraframe. The temporal redundancies of a frame sequence can only be used if differential frames are transmitted. So it is also advisable to use subband coding for differential frames. A further improvement can be achieved by using subband coding for a motion-compensated temporal DPCM.

The main advantage of subband coding as compared to transform coding is the better use of the frame redundancies, since the word lengths of the subbands are greater than the typical block size in transform coding.

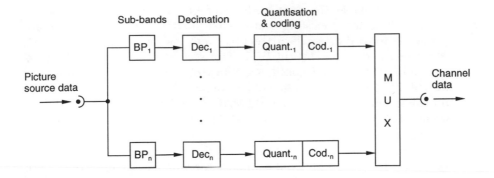

Fig. 3.53 Principle of subband coding with scalar quantisation [185]

Moreover, the block effects from transform coding can thus be avoided. Subband coding also provides a better support for the use of hierarchical picture transmission. It has, however, drawbacks, such as edge crawling caused by the ringing of the bandpass filters, and the cost and complexity of the receiver [185].

3.7.2.9 Object coding

This coding concept claims to recognise objects which are transmitted in the form of an addressed scene description, causing the most drastic bit rate reduction. With object coding, the picture sequence is no longer understood as an arrangement of statistically dependent pixels. The object coder sees a three-dimensional scene of mobile and immobile objects with certain characteristics such as shape and colour. Obviously, only a small data quantity is required to describe the analysed picture sequence, but a very complex catalogue of objects. Object coding can be seen as an expert system — a category of artificial intelligence. It is so far evident that in view of the immense information bank on objects required and the complex algorithms of recognition the application of object coding will be limited at present to only simple pictures, such as the video telephone [186].

3.7.2.10 Standards

In connection with video coding, there are several standards organisations, three of which are mentioned below:

- ISO develops industry standards on which it advises the United Nations. From 1991, ISO and IEC (International Electrotechnical Commission) have been co-ordinating work in the field of telecommunications with the JTC1 (Joint Technical Committee 1).
  MPEG (motion pictures expert group) is a subdivision of JTC1 for the definition of a standard for motion picture communication. This standard

should be suitable for multimedia storage on CD-ROM but also be compatible with the existing transmission media. The MPEG-1 standard is suitable for coding small-size pictures with low data rates (1.5 Mbit/s). The video coder uses hybrid DCT. The second project phase MPEG-2 started in 1992. MPEG-2 specifies a compatible method that is preferably with MPEG-1 and allows coding of an enhanced PAL quality with 4 Mbit/s providing quality better than D2-MAC with 10 Mbit/s. The MPEG-2 standard also includes HDTV up to 40 Mbit/s. MPEG2 followes a generic coding concept, which defines the syntax of an algorithm satisfying the different application requirements with the appropriate data rates [167, 182, 183, 184]. The MPEG-2 standard for video baseband coding was passed as a Committee Draft in November 1993 in Seoul, Korea, to become the standard by late 1994. Similar to audio baseband coding, MPEG-2 seems to be emerging as a world standard for video, both at the transmitter end and in the receiver at the demodulator output: OFDM for terrestrial reception, PSK for satellite and QAM for cable reception.

- EBU, the union of European broadcasters, is standardising sound and TV broadcasting while a Joint Technical Committee with ETSI is to ensure its harmonisation.
- ETSI has been established for the definition of European standards, ETS (European Telecommunications Standard).

Regarding the scientific research of data compression using psycho-optical effects, video coder technology is in a phase where the findings have been specified in the form of algorithms, creating a basis for chip technology to be used. VLSI chips have for instance been developed for DCT and for motion estimation. Systems implemented on video processors are also already available, so it is expected that in 1995 video coding will be industrially available for high- to limited-quality TV [180, 188].

3.8 Digital HDTV

3.8.1 Transmission chain

Fig. 3.54 shows the transmission chain for digital HDTV broadcasting. The high-quality video input signal may be taken via a signal processing unit before being applied to the source coder. The output data rate of 30 to 20 Mbit/s is subjected to FEC (forward error correction) channel coding. In the subsequent multiplexer, the audio and sync. data and the control signals are added to the coded picture signal, whereupon a gross data rate of 30 to 40 Mbit/s is obtained. The channel-coded data are modulated onto the IF carrier and packed into a channel bandwidth of 7 or 8 MHz (6 MHz in the US). A mixer upconverts to the RF carrier frequency, e..g in Band I to Band V (47 MHz to 862 MHz), before the signals are applied to the terrestrial transmitter with power amplifier and transmitting antenna.

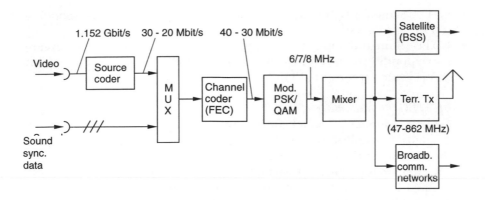

Fig. 3.54 Transmission chain for digital HDTV

In the receiver, signal processing is basically reversed to that of the transmission with the signal received either via satellites or broadband communication networks. Due to the different channel transmission characteristics of the various media, channel coding and modulation techniques are adapted accordingly. The counterargument is of course that it should be possible to connect the receiver to any of the media without any additional modification.

In Europe, the concept of a hierarchical quality structure is being followed for the digital TV picture signal (Table 3.10):

Quality classification	HDTV High	EDTV Enhanced	SDTV Standard	LDTV Limited
comparable to	2 x CCIR601	CCIR601	PAL SECAM NTSC	VHS
Approx. data rate after source coding	30 Mbit/s	11 Mbit/s	4.5 Mbit/s	1.5 Mbit/s (MPEG1)

Table 3.10 Hierarchical quality structure for video

- HDTV describes a quality level that is comparable to the HDTV studio standard (better than HD-MAC).
- EDTV (enhanced definition TV) quality is comparable to the studio standard following CCIR Recommendation 601.

- SDTV (Standard Definition TV) means a quality standard corresponding to PAL or SECAM.
- LDTV (limited definition TV) defines a standard for mobile reception corresponding to the MPEG-1 system and is comparable in picture quality to VHS video recorders [193].

3.8.1.1 FEC channel coding

Error protection and channel coding are used to protect the source-coded data against transmission errors. The degree of protection has to be adapted to the error characteristic of the transmission channel. Block codes, such as the Reed-Solomon codes, are suitable for this error-correcting coding. RS (224, 208) for instance is an algorithm allowing correction of up to 8 byte errors or 58 continuous bit errors in a code word. The coder adds 16 redundant checkbytes as code words to each block consisting of 208 data bytes. CRC-based error detection is used in addition where transmission errors occur at such a frequency that the capacity of the Reed-Solomon code is insufficient for error correction. CRC-based error detection allows error concealment in destination decoding [189].

3.8.1.2 Avoiding abrupt degradation of picture quality

It is considered to be an advantage of analogue signal transmission that with increasing distance to the transmitter there is a linear degradation of the picture and sound quality even when additional adverse effects are caused by the topography. With a digital transmitter on the other hand the signal is received with constantly good quality as long as there is sufficient field strength at the receiver to allow almost error-free decoding of the digital datastream in the demodulator. If the receiving field strength falls below a certain minimum it causes an abrupt increase in the number of bit errors, failure of the frame synchronisation and consequent total failure of the signal decoder (Fig. 3.55). This effect is found to be annoying by the viewer and is best avoided by a graceful degradation of signal or picture quality similar to that in an analogue transmission.

The problem can be solved by channel coding. A feasible approach is, for instance, the definition of data packets of constant length with the number of error protection bits to be adapted to the respective priority class (Fig. 3.56). This would allow the definition of a data structure of constant-length frame data, as used in ATM (asynchronous transfer mode) networks [189].

Another method of implementing graceful degradation is the introduction of a data hierarchy in the source coder with individually variable error protection. Different priorities are assigned to the various information contents of the source coder to be transmitted. If the receiving field strength drops below the minimum level at the corresponding distance from the transmitter, the data having lowest priority will become first unusable and are not considered in the decoding. Consequently an HDTV

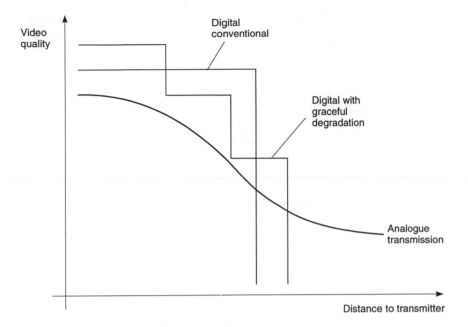

Fig. 3.55 Graceful degradation

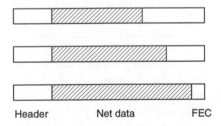

Header   Net data   FEC

Fig. 3.56 Possible data structure with constant cell length for graceful degradation

picture will be decoded but with reduced quality and a further decrease in the system-inherent sensitivity with the data of the next priority class becoming unusable and causing a further degradation of picture quality.

3.8.1.3 Multiplexer
DSP (digital signal processors) are used for formatting and combining the video, audio and control data into a single datastream. The first function is performed by the video multiplexer which combines the data from the buffers of the source coder with the sync. frame and headers. The second

function is performed by the multiplexer, which adds the audio and control data channels to the video datastream and superimposes its own header, clock and sync. information.

3.8.1.4 Modulation

The source- and channel-coded video and audio signals are taken via a suitable programme line to the terrestrial transmitter where the digital signals are modulated onto one or several carriers with the gross data rate transmitted. For selecting the appropriate modulation method, a high bandwidth efficiency is required on the one hand, and a low susceptibility to interfering signals on the other. Generally it can be said that the greater the modulation depth, i.e. the more modulation points within the four quadrants of the co-ordinate system, the higher the bandwidth efficiency, i.e. the more data bits are coded in a modulation symbol. It should be considered that the closer the modulation points within the co-ordinate system, the smaller is the decision range in the demodulator, and the lower is the transmission system's immunity to interference.

Basically, there are two modulation methods for digital television:

- In the US, single- or dual-carrier methods are mainly used and provided with high-order quadrature amplitude modulation.
- In Europe, an OFDM method based on the method used in DAB according to EUREKA 147 is generally used in the various techniques for terrestrial digital HDTV.

With single-carrier QAM, at least 16-QAM has to be used to achieve a gross data rate of at least 20 Mbit/s for a spectral efficiency of theoretically 4 bit/s per Hz and relative to the data rate that can actually be transmitted, 32- or 64-QAM with 5 or 6 bit/s per Hz being used. The modulation depth is limited by the channel susceptibility to interference and the multipath propagation in terrestrial broadcasting. A good compromise is 32-QAM, allowing a gross data rate of about 30 Mbit/s in a 7 MHz channel.

A special version of the single-carrier QAM is the dual-carrier QAM (Twin QAM), which leaves gaps in the digital modulation signal in that component of the TV channels, in which, for instance, a PAL TV signal has high energy components in the range of the vision carrier, the colour subcarrier or the sound carrier. These frequency gaps in the digitally modulated HDTV signal cause a high degree of compatibility between TV channels with conventional modulation methods and channels with digital modulation.

The OFDM method features special advantages in terrestrial multipath propagation. By distributing the gross data rate among many, for instance 1000, carriers, the data rate per carrier is low and — reciprocally — the bit duration high. Through the additional insertion of a guard interval, echo signals can be utilised so they contribute towards the useful field strength (Section 2.5). Additional interleaving in the time and frequency domain ensures that signal fading of individual OFDM carriers will have no adverse effect. OFDM also provides the possibility of implementing single-frequency networks featuring high frequency economy.

3.8.2 Digital HDTV in the US

In the US there is a very special situation between two used 6 MHz channels with one channel always left free, the so-called taboo channel, to avoid interference in the adjacent channel-compatible receivers. In view of the compatibility to be expected between digitally modulated carriers and TV channels with conventional modulation modes, it was decided to release the taboo channels for terrestrial simultaneous broadcasting (simulcast) both in NTSC and in the ATV (advanced TV) quality. For developing the appropriate broadcasting system, FCC (Federal Communications Commission) initiated ACATS (Advisory Committee on Advanced Television Service). In co-operation with the receiver industry, the broadcasters established ATTC (Advanced Television Test Center). The proposals made by the several competitors were all digital, with the exception of the Japanese Narrow MUSE, being withdrawn early in 1993. The FCC expected the final decision on a system or a combination of several systems (Grand Alliance) to be made in 1994 along with the published specifications.

Future plans are for HDTV to be in operation by the year 2000 with NTSC to be withdrawn from 2009.

The five HDTV systems originally proposed to the ATTC were:

- Narrow MUSE by the Japanese broadcasting corporation NHK
- DigiCipher by General Instrument Corporation (GI)
- Digital Spectrum Compatible HDTV (DSC-HDTV) by Zenith/AT&T
- Advanced Digital HDTV (AD-HDTV) by Advanced Television Research Consortium (ATRC)
- Channel-Compatible DigiCipher HDTV by Advanced Television Alliance (MIT/GI).

The characteristic data of the five systems are given in Table 3.11 [196].

3.8.2.1 DigiCipher

The DigiCipher HDTV system is an integrated system for high definition digital television, digital audio with CD quality and data and text services, to be transmitted via a VHF or UHF TV channel. An additional function for conditional access, i.e. pay TV, is also provided, with encrypted video, audio and data services (Fig. 3.57).

A typical system of hybrid DCT is used for baseband coding. Usually, an interframe mode is employed, the two fields being added to give a frame. A special feature of this code is the regular transmission of all picture components with pure intracoding. This is necessary as at the beginning of the transmission or after transmission errors the predictor in the receiver

	Narrow muse	DigiCipher	DSC-HDTV	AD-HDTV	CC-DigiCipher
Transmitted scan lines	750	960	720	960	720
Sampling rate Y (MHz)	40.1	53.7	75.3	56.6	75.3
Encoder input data rate (Mbit/s)	246.3	405.1	745.7 (9 bits per pixel)	517.9	1,325.8
Compressed video data rate (w/o FEC)	4.86 MHz analogue baseband	18.24 Mbit/s (13.34 Mbit/s with 16 QAM option)	8.46–16.92 Mbit/s variable rate	17.73 Mbit/s	19.26 Mbit/s (estimated)
Compression ratio	5.5 : 1	22.2 : 1	44.1–88.2 : 1	29.2 : 1	68.9 : 1
Modulation	Split-channel VSB-AM/ SSB-AM	32 QAM (16 QAM option)	Dispersed 4-VSB/2-VSB	Spectrally shaped QAM (32 QAM	32 QAM (16 QAM option)

Table 3.11 Basic parameters of US proposals for HDTV/EDTV

contains random data. With pure intracoding, the receiver predictor is synchronised with the transmitter predictor. The sequence of the intracoded transmission determines the time taken for reconstructing the picture (acquisition time).

The frequent switchover to other TV channels via remote control is problematical. DigiCipher therefore uses intracoding for each block every 11 frames with time interleaving.

Motion compensation is based on a block size of 16 × 32 pixels. The search area is ±32 horizontally and ±8 vertically with an accuracy of 1 pixel.

To reduce with the DigiCipher coder the input data rate of about 1 Gbit/s to the output data rate of about 13 Mbit/s, the U and V chrominance signals are prefiltered and subsampled by a factor of 4 horizontally and a factor of 2 vertically. The additionally required compression factor of 30 can just be provided by hybrid DCT.

In addition to the 16-QAM method, DigiCipher also offers the choice of 32-QAM with 24.5 Mbit/s gross channel data rate. Due to the more demanding C/N requirements, the greater modulation depth for the same transmitter power reduces the coverage area. To put it another way, the better the picture quality the less the coverage area, resulting in a reduced number of viewers.

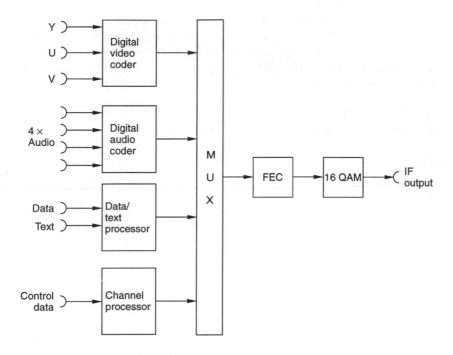

Fig. 3.57 DigiCipher coder/modulator

An inner error protection is implemented by means of a convolutional code, which is decoded by a soft-decision trellis decoder, and an outer error protection by Reed-Solomon coding. Burst errors are overcome by twofold interleaving.

For satellite broadcasting, DigiCipher uses a single-carrier method with time multiplex structure allowing the transponder to be operated close to saturation point, so that the transponder output power is optimised with the minimum size of parabolic receiving antenna. Satellite transmission makes use of QPSK modulation with 27 Mbit/s, requiring a satellite transponder of 27 MHz or more. For a bandwidth of 27 MHz the DigiCipher satellite system needs a carrier-to-noise ratio of only 7 dB, which is better than, for example, a 20 dB C/N for analogue or hybrid HDTV systems.

For terrestrial broadcasting, the DigiCipher system employs QAM in two modes. 16-QAM with 4.88 MSPS (mega symbols per second) is used in areas with common-channel and adjacent-channel interference, whereas 32-QAM with 4.88 MSPSl is chosen where interference is noncritical. The receiving threshold is 13 dB C/N for 16-QAM and 17 dB C/N for 32-QAM.

An adaptive echo equaliser is incorporated in the receiver to compensate for interference such as reflections in the multipath reception and group-delay distortion caused by transmitter amplifier antennas or tuners.

Cable transmission allows 64-QAM with a symbol rate of 4.88 MSPS with the cable system carrying the full data capacity of the DigiCipher satellite system in a single 6 MHz cable channel with a C/N receiving threshold of 28 dB. A cable channel is also able to transmit the same data capacity as a satellite transponder [194, 197, 198].

3.8.2.2 DSC-HDTV

DSC-HDTV has been jointly developed by Zenith and AT&T. The designation DSC (digital spectrum-compatible), represents a technique in which four digital audio channels in CD quality and ancillary services in the form of digital data (e.g. teletext), are transmitted in addition to the HDTV signal in a 6 MHz channel (Fig. 3.58).

The DSC video coder uses hybrid transform coding. A special feature of this coding method, is, for example, hierarchical block matching, where first coarse and fine vectors are determined for motion compensation. To adhere to the bit capacity for vectors, there is a selection algorithm for vectors to be transmitted with fine resolution and those to be transmitted with coarse resolution. The coder compression reduces the data rate of 1.5 Gbit/s at the input to 17.2 Mbit/s at the output to which ancillary services and FEC are added.

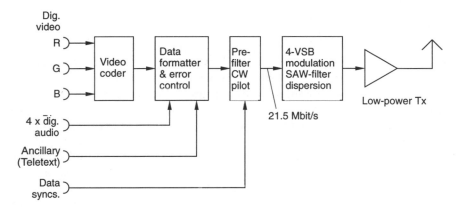

Fig. 3.58 DSC-HDTV transmitter

The gross data rate of 21.5 Mbit/s is modulated onto the transmission band using four-level ASK (amplitude shift keying). The lower sideband is suppressed by a special residual sideband filter to obtain a 6 MHz vestigial sideband spectrum. The choice was made favouring VSB (vestigial sideband) modulation because of its excellent compatibility with NTSC channel signals (Fig. 3.59).

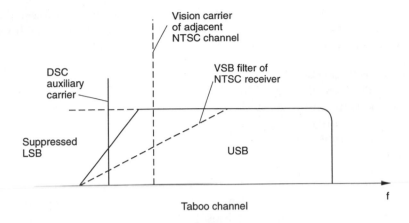

Fig. 3.59 VSB spectrum of DSC-HDTV method

In the DSC system, the four-level modulation can be switched to a 2 VSB ASK for high-priority information. Since these data are transmitted with a high signal power/bit, decoding in fringe areas of the coverage area still gives acceptable picture quality (Section 3.8.1.2) [199].

3.8.2.3 ADTV-MPEG

The ADTV system was proposed by ATRC (advanced television research consortium), formed by various research centres and institutes. For compression, an MPEG-1 coder based on line interlacing is used. This method is therefore referred to as ADTV/MPEG++. Graceful degradation is ensured by priority data being transmitted with high-order QAM. The data rate is caused by a spectrum which fits into the window below the NTSC vision carrier. The information in the lower QAM band (Fig. 3.60) is transmitted with a higher power per symbol and can therefore be reliably decoded in the fringe areas.

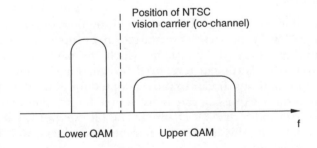

Fig. 3.60 Twin QAM in ADTV/MPEG++ method

The upper QAM band transmits the low-priority data with lower power per symbol. The critical region of the NTSC vision carrier is not used for data transmission, bringing into use the name 'twin QAM' [180].

3.8.2.4 CC-DigiCipher

This method was developed jointly by GI (General Instruments) and MIT (Massachusetts Institute of Technology). It uses progressive sampling, without line interlacing. The method is also referred to as Channel-Compatible DigiCipher and as the ADTV-P (American Progressive HDTV) system.

3.8.2.5 DirecTV

In December 1993, the USA launched their first 2.8 ton direct broadcast satellite DirecTV for digital television [249].

DirecTV is equipped with 16 transponders of 120 W each, making it the most powerful commercial satellite in the world. DirecTV, the programme providers, uses 11 transponders and the US Satellite Broadcasting Inc., of the Hubbard group uses five.

Subscribers will eventually be able to receive 75 digital TV programmes — 4 or 5 per transponder — with a 46 cm dish and input filter from Thomson-RCA at a cost of 500 US$ (£200). A second satellite of similar capacity is to follow in 1995.

The service is provided on a pay-per-view basis, with the subscriber paying a monthly subscription by a smart card for the programmes seen. The studio is a reproduction centre in Castle Rock equipped with 300 Digital Betacam Systems from Sony and the satellites built by Hughes Space & Comm. Inc., Los Angeles, a General Motors subsidiary.

The core of the DirecTV transmission technique is a cascadable superchip for compression and decoding of the digital high-definition TV signal, developed by the high-tech companies Compression Labs and C-Cube Microsystems.

3.8.3 Digital TV in Europe

Following up the groundwork prepared by the US in the field of terrestrial digital HDTV, various HDTV projects were started in Europe one or two years ago. Unlike the American HDTV systems, which are mainly based on single-carrier methods, the European terrestrial broadcasting systems use exclusively multicarrier methods similar to those adopted for DAB. With the European OFDM method, the gross bit rate is distributed for instance among 1024 carriers in the bandwidth of a TV channel. The bit rate per carrier is thus the total gross data rate divided by the number of carriers, with the following condition applying:

$B/N=1/T_{useful}$

(B: bandwidth, N: number of carriers, $t_{Symb}= T_{useful} + T_{guard\ interval}$)

Depending on the modulation depth selected, two bits can be transmitted during the symbol period with QPSK, four bits with 16QAM, five bits with 32QAM and six bits with 64QAM.

The maximum bit rate transmitted within a given bandwidth is determined by the laws of physics. In this respect OFDM methods and QAM single-carrier methods are equivalent, since the maximum bit rate to be transmitted is determined by the noise power in the channel.

There are other reasons giving an advantage for OFDM: the low bit rate per carrier results in a long bit period allowing insertion of a guard interval during which the receiver does not evaluate the transmitted signal. If the guard interval is longer than the maximum echo delay times occurring in the transmission channel, reflected signals contribute towards the useful signals. For DAB, a guard interval of 25% of the useful bit period is selected. Multipath reception may only cause amplitude and phase errors of the individual carriers which are largely eliminated by time and frequency interleaving of the signal information.

Another advantage of the guard interval method is the possible implementation of SFN (single-frequency networks). The SFN transmitters are frequency-coupled and send exactly the same bits at any given time. The SFN allows nationwide terrestrial distribution of the HDTV signal in a TV channel. The frequency resource can thus be utilised to a far better extent than with conventional frequency planning (improvement factor of about 3). Moreover, the ERP required by the terrestrial digital transmitters is approximately ten times lower than that of conventional analogue TV transmitters. Using the OFDM method, 12 to 30 Mbit/s can be transmitted in an 8 MHz TV channel, depending on whether stationary, portable or mobile reception is required.

3.8.3.1 HD-DIVINE

The Swedish Telecom and Swedish broadcasters are closely co-operating with the Telecoms of Denmark and Norway to develop a terrestrial digital HDTV prototype system. The project has been given the name HD-DIVINE (digital video narrowband emission). The main purpose of this project is to contribute in the course of the 1990s to a terrestrial digital HDTV standard. Its chief goal is to demonstrate the feasibility of digital HDTV broadcasting in an 8 MHz TV channel. At the 1992 IBC (International Broadcasting Convention) in Amsterdam an initial hardware concept was implemented.

The system is based on the HDTV studio signal according to the European studio standard of 1250 lines/50 Hz/2:1. The video signal is compressed to 24 Mbit/s by using motion-compensated hybrid coding. The coding algorithm is based on the CCITT (H261), ISO/MPEG and CCIR/ETSI standards. Further investigations within the HD-DIVINE project include modern motion detection/compression and the compression of motion vectors through the use of intra-/inter-DCT coding [192]. The total system consists of a video coder and a decoder with an interface to the OFDM modem for the transmission of 27 Mbit/s – including FEC – in terrestrial 8 MHz channels (Fig. 3.61). Four audio channels of 128 Kbit/s each are transmitted in mono mode and the audio signals are coded according to the ISO Layer-II model (MUSICAM). A data channel of 64 Kbit/s is also provided. Together with the 2 Mbit/s FEC (RS 224, 208) a gross data rate of 27 Mbit/s is obtained. A special feature of the codec is that bit rates between 15 and 45 Mbit/s can be selected.

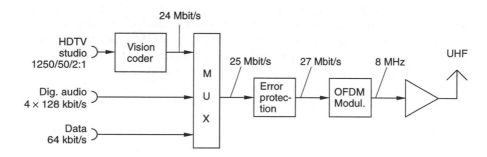

Fig. 3.61 HD-DIVINE prototype system

The following reasons are given for choosing the OFDM method: with a single-carrier method and a transmission bandwidth of 8 MHz the symbol time can be expressed as:

$$T_S = 1/(8 \times 10^6 \text{ Hz}) = 0.125 \text{ µs}$$

A 512 carrier OFDM features a symbol period that is 512 times longer, namely 64 µs. In the case of a reflection of for instance 1 µs, the OFDM symbol period is much longer than the reflection period. With a single-carrier method, the situation is exactly the opposite.

To avoid intersymbol interference, the symbol period is extended by the guard interval, which is 2 µs in the prototype system. The HD-DIVINE system is based on stationary reception, i.e. there is no significant change to the transmission channel during symbol transmission [200].

3.8.3.2 STERNE

The STERNE (Système de Télévision en Radiodiffusion Numérique) project is mainly supported by the French research institute CCETT. The STERNE project specifies and develops prototype equipment for terrestrial digital TV broadcasting on the high performance video coding and adaptation of OFDM to TV requirements. The main objective of the CCETT STERNE project is terrestrial HDTV broadcasting for the stationary receiver. The channel bandwidth is 8 MHz. The quality level of picture and sound should correspond to that of the D2MAC/packet. Access coding, the transmission of programme-associated data and multiplexing of several programmes in a single channel are further objectives of this project.

The video coder is based on hybrid DCT, the algorithm of which has been developed by CCETT, and on an already developed codec for programme transmission with 34 Mbit/s.

The STERNE video coder uses a coding rate of 5 Mbit/s for SECAM quality and of about 10 Mbit/s for 4:2:2 studio quality. For HDTV quality, a coding rate of 30 Mbit/s is required.

The STERNE project has adapted the OFDM method used in DAB to the requirements of TV broadcasting. The datastream is distributed among a large number of narrowband carriers. Channel coding using a convolutional code in conjunction with a decoding method based on maximum likelihood ensures error-free decoding even if several individual carriers of the multicarrier band are distorted. Statistical independence of the signal elements is achieved by interleaving the information in the time and frequency domain, eliminating the effect of Rayleigh channel fading. The STERNE project includes an access control according to the Eurocrypt standard.

Demonstration of a broadcasting service with high data rate to stationary receivers, both in HDTV and in conventional TV quality, was scheduled for the end of 1993. By 1999, the STERNE project intends to ensure that terrestrial broadcasting will distinctly improve the picture quality for portable receivers [201].

3.8.3.3 Digital video broadcasting DVB by Thomson/WDR

In 1992 the French Thomson group, in co-operation with the German University of Wuppertal, installed digital signal coders and transmission equipment in the transmitter station Langenberg of the WDR broadcasting corporation. The receiving unit underwent both stationary and mobile testing [202].

The output datastream of 216 Mbit/s was provided by a D1 recorder while a commercially available digital DCT coder with additional motion compensation and differential frame coding according to the ETSI standard was used for picture coding. The digital TV signal compressed to 34 Mbit/s is OFDM modulated onto 512 subcarriers with 6 bits per sample (Fig. 3.62).

This pilot project uses channel 38 which is normally reserved for radio astronomy. The amplifier supplies an output power of 6 W with the antenna vertically polarised. Further system data are:

- useful bit period 70.4 µs
- guard interval 8.8 µs
- carrier difference 14.205 kHz

- number of carriers 512 (491 usable)
- spectral bandwidth 7.06 MHz
- a test symbol is inserted after every 15 symbols.

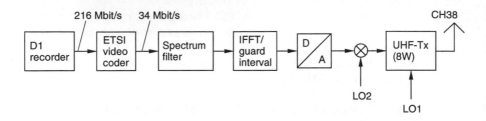

Fig. 3.62 HD digital test system by Thomson-CSF/LER

3.8.3.4 dTTb

The goal of the dTTb (digital terrestrial television broadcast) project is the investigation of terrestrial TV broadcasting in the UHF and VHF band, with a 16:9 aspect ratio for stationary reception, but also for portable and mobile reception. It is emphasised that the situation of European broadcasters is quite different to that in the USA and this may lead to the implementation of a system with different specifications taking account of the different requirements. A large number of European research institutes, broadcasters, the consumer electronics industry and manufacturers of professional system equipment are participating in the dTTB project.

The aim of the project is hierarchical picture coding. The two-dimensional picture is divided by a filter bank into a number of subbands each of which is individually coded (Fig. 3.63).

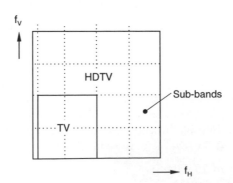

Fig. 3.63 Principle of hierarchical subband decoding

For channel coding and modulation, both single-carrier systems in conjunction with adaptive equalisation and multicarrier methods according to the OFDM model are used. For carrier modulation, QPSK, 8PSK, 16PSK, 16QAM and VSB-QPSK are being considered [204].

3.8.3.5 HDTV-T

The Federal German Ministry for Research and Technology initiated the HHI project (Heinrich-Hertz-Institut) for work on the terrestrial digital HDTV broadcasting definition phase which was started early in 1991. Representatives from consumer and capital goods industry and research institutes and universities are participating in this German HDTV project. The project deals with the fundamentals of HDTV coding with 20 to 30 Mbit/s, the modulation methods and terrestrial digital broadcasting and with aspects of network operation. By picture scaling, compatible broadcasting of TV/HDTV with different quality standards is to be initiated.

The source coder reduces the HDTV video signal to approximately 20 Mbit/s. Together with the digital sound signals and the synchronisation signals as well as the FEC coder, a datastream of approximately 30 Mbit/s is obtained at the output of the multiplexer. The digital bitstream must be transmitted in a bandwidth of 7 to 8 MHz by the RF modulator. Project work also includes the definition of joint interfaces with the various technical media. According to the time schedule of this national co-operative project HDTV-T, the hardware implementation should be completed by mid-1995 followed by field trials for system optimisation and specification [205, 206].

3.8.3.6 Vidinet

The Vidinet project (video in digital networks) is carried out by German DBP-Telekom/FTZ Berlin with the aim of demonstrating a single-frequency network with five transmitters at the International Consumer Electronics Show 1993 in Berlin. The original intention was to use a generic code for the picture quality (Fig. 3.64).

In a trial operation, an NTL baseband coder was used for picture coding and an NTL multiplexer for four programs of 6 Mbit/s each. Together with a concatenated Reed-Solomon error protection, a total data rate of 34.368 Mbit/s is obtained.

The system uses a uniform 64QAM-OFDM mode designated as 64 DAPSK (differential amplitude PSK). Classical 64QAM with equidistant uniform modulation points is to be replaced later by a nonuniform 64QAM with different distances between the modulation points within the four quadrants.

This pilot project features the following further system parameters: graceful degradation, transmitter distance 28 km, transmitter power 0.3 to 0.5 kW (approximately 4 kW ERP), receiving field strength 75 dBμV/m, transmitter frequency channel 59 (alternatively 61), 1024 carriers (1-k FFT), guard interval approximately 20 μs. A special feature of this pilot project to note is that nonlinear distortions are considered in channel simulation, in particular because of the 64QAM mode. The experimental phase is scheduled to last until 1995 and the system specification is expected to be completed by 1997. This new technique will also support interactive television and multimedia.

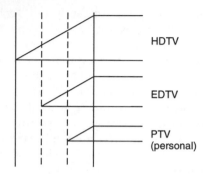

Fig. 3.64 Principle of generic code as used in DBP-Telekom project Vidinet

3.8.3.7 DIAMOND
Thomson-CSF/LER has developed a video codec for HDTV with 34 Mbit/s in co-operation with the BBC (British Broadcasting Corporation) for conventional TV with 8 Mbit/s (ETSI-300174 algorithm). At the International Symposium and Exhibition in Montreux in June 1993 a broadcasting system based on this concept was demonstrated in UHF channel 43, transmitting 34 Mbit/s with horizontal and vertical polarisation of the transmitting antennas. The system uses OFDM with as many as 512 carriers and 64 QAM. For the demonstration, HDTV signals from a D1 tape recorder source and four TV programmes were broadcast.

The aim of the DIAMOND project is to implement multilayer channel coding in conjunction with hierarchical source coding, versatile signal multiplex, conditional access and connection to ATM broadcast networks.

3.8.3.8 SPECTRE
The SPECTRE (special purpose extra channels for terrestrial radiocommunication enhancements) project was already published at the IBC 1990 and field trials, at the time of the IBC 92, were carried out at Stockland Hill and Beacon Hill in Devon, UK [203].

The video coder employed compresses the data rate from 216 Mbit/s to 12 Mbit/s using predictive and interpolative HDCT. The multiplex signal is protected by a burst error correction code RS (255, 239) so that 13 Mbit/s are taken to the OFDM modulator. Approximately 400 carriers are QPSK- or 8PSK-modulated. In the transmitter unit the multicarrier signal is converted into the UHF channel and boosted via two linear 200 W tube transmitters. The signal is radiated via a log-periodic antenna on a 110 m mast (gain 8 dB, feed loss 3 dB). The SPECTRE transmitter power is 250 W ERP, being 30 dB below the 250 kW ERP of the Stockland Hill PAL transmitter. The SPECTRE system parameters are flexible according to specification and for the prototype equipment, allowing, for example, 24-Mbit/s coding or 16PSK/16QAM.

The introduction strategy followed in the UK is to digitally simulcast existing programmes with low power in the taboo channels and with conventional power to the PAL-I standard.

3.8.3.9 Digital television on ASTRA

SES-ASTRA is implementing an ambitious project on the introduction of digital television on transponders D, E and F (Section 1.2.3, Table 1.2).

The basic specification has to meet the requirements of all participants in broadcasting, ensure maximum compatibility with the technical media — satellite, cable and terrestrial transmitters — and to meet the competitive requirements of the market in the form of conditional access.

In addition to the picture quality (VHS through to studio quality, 4:3 and 16:9) programme-related information is to increase the benefit to the subscribers, with the provision of more variety and choice of programmes, according to individual criteria with real-time decisions, seen as essential.

Other keywords are:
- pay TV
- pay-per-view
- (near) video-on-demand.

SES-ASTRA takes an active part in the preparation of the system specification in the various committees: MPEG, European DVB project (Section 3.8.3.10), ETSI and CCIR.

The standard is based on MPEG-2 for video source coding, audio source coding (MUSICAM) and signal multiplexing.

QPSK has been chosen for satellite broadcasting and 64QAM for cable networks.

The main BSS transmission parameters are:

- transponder bandwidth, nominal   33 MHz
- gross data rate   approximately 55 Mbit/s
- net information data rate   30 to 40 Mbit/s
  depending on FEC.

A concatenated FEC code (convolutional code plus block code) is employed for channel coding.

Depending on the picture quality transmitted, a transponder has the following typical channel capacity:

- films   2 Mbit/s   12 to 16 channels
- sport   6 Mbit/s   5 to 6 channels
- EDTV   9 Mbit/s   3 to 4 channels

Digital TV broadcasting via ASTRA 1E using available receiving systems is scheduled for mid-1995.

3.8.3.10 European digital video broadcasting project DVB

The representatives in the DVB Launching Group, as it was initially called, is a European co-ordination group, from industry, public and private broadcasters, network operators, administrative authorities, research institutes and the EU commission. The aim of the co-ordination group was to prepare by 1993 a proposal for digital video broadcasting via cable and satellite taking into account the various European interests. Implementation and market introduction of the European DVB system was scheduled for 1995.

In September 1993 the European DVB project was inaugurated upon presentation of a Memorandum of Understanding, which by the end of 1993 was accepted by about 100 signatories from 12 countries.

In December 1993 a baseline specification was already submitted for a European system for digital TV, sound broadcasting and data services under the working title: The MULTIVISION System [248].

Broadcasting is made via satellite in the 11/12 GHz band and via CATV cable, as well as via SMATV (Satellite Master Antenna Television).

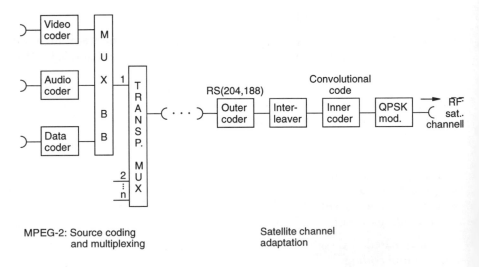

Fig. 3.65 Block diagram of DVB baseline system

The baseline system contains the functional blocks, from the baseband (BB) and MPEG-2 multiplexer through to satellite adaptation (Fig. 3.65).

The integration of the various European projects in a system solution for all technical media jointly supported by all European and some other countries would be a great success for the European DVB project.

Chapter 4

Pragmatic models for future developments in digital sound and TV broadcasting

This chapter deals with the aspects of future sound and TV broadcasting by looking at previous and present-day developments. These pragmatic approaches, as described in Chapters 2 and 3, are based on methods which are either tried and tested, or are in the process of being specified. The combination and enhancement of these methods gives systems improved quality and performance, due to modification and improvement of the existing hardware.

4.1 DSRplus

DSRplus is an improved DSR method specialising in the fields of application of digital satellite and cable broadcasting.

In Germany, DSR is broadcasting on the satellites and gaining popularity on TV-SAT2 and KOPERNIKUS and on the cable channels S2 and S3 at 118 ±7 MHz [210]. In Switzerland, a community antenna is fed via a microwave link feeding into the cable network feeding network. While other European countries are deciding on the introduction of DSR [212], the receiver industry is preparing to promote DSR by pricing the receivers attractively.

Currently DSR has the following disadvantages:

- a capacity of only 16 stereo programmes
- practically no data reduction
- no particular frequency economy, since the gross data rate of 20.48 Mbit/s is distributed via a 27 MHz transponder or in a cable of 14 MHz bandwidth
- the introduction of the system was too long, due to the loss of the TV-SAT1 satellite
- the concept of the DSR method was not compatible with European requirements, so delaying its introduction in other countries.

The DSRplus method proposed below, eliminates these advantages while being compatible with the current DSR method [207, 208].

4.1.1 Combination of DSR and MUSICAM (= DSRplus)

To meet the requirements of compatibility, the net capacity of the digital sound transmission channels in DS1/DSR is used to transmit several programmes per DS1 by using, for instance, MUSICAM, instead of just one stereo signal per DS1 line (Fig. 4.1). The insertion into the DS1 frame is very easy since the DS1 interface is a 2 × 16 bit parallel interface with 32 kHz sampling rate providing transparent transmission of 2 × 14 bits, i.e. output bit pattern = input bit pattern. The scale factor employed in DS1 and the resulting bit interleaving can be maintained because of the high transmission reliability of the scale factor. By defining an appropriate scale factor, bit transmission without interleaving can also be implemented. Parallel inputs for 2 × 14 bit/32 kHz are provided on the DS1 line transmitter which can easily be adapted to an integral multiple of the 32 kHz clock rate. The clock rates then directly correspond to the bit rates defined by MUSICAM, e.g. 64.96 and 128 Kbit/s per mono signal [213].

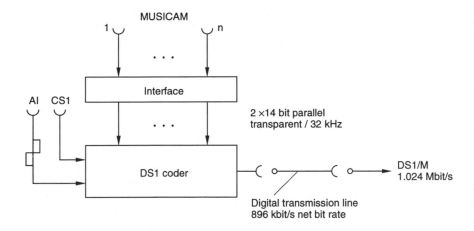

Fig. 4.1 MUSICAM/DS1 interface

4.1.2 DSRplus transmitter system

The DS1 channels occupied by MUSICAM signals are routed to the satellite transmitter in the earth station (Fig. 4.2). Through the existing line network, MUSICAM can either be transmitted on, for example, ISDN or DS1 signals [209].

For the application of the DSRplus method, the concealment function of the satellite transmitter on the DS1 lines according to DS1 specification has to be disabled. The 16 input signals are then multiplexed in line meeting the DSR standard.

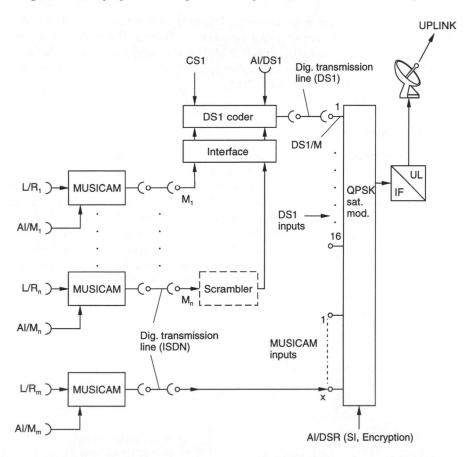

Fig. 4.2 DSRplus system: from studio to earth station with MUSICAM via DS1 or directly to satellite modulation

DS1 coders with the associated interfaces for the MUSICAM signals may be omitted if the multiplexer/QPSK modulator is provided with modules converting the MUSICAM signals at the input directly into the DSR frame structure.

Regarding the DSR frame, note that the block coding in the DSR channel distinguishes between a protected and unprotected bit transmission. Protected means that two bit errors per BCH block can be corrected. Due to the correction, the C/N ratio is improved by approximately 4 dB in the case of a corresponding bit error rate.

In the DSR channel, the following bits are available per DS1:

2×11 bit/32 kHz DSR protected

2×3 bit/32 kHz DSR unprotected.

This differentiation may, however, be disregarded since the MUSICAM baseband coding has its own error protection. With a BER of less than 10^{-3} the subjective quality impairment of MUSICAM is negligible. Since this BER condition applies to satellite and cable in DSR, the three DSR unprotected bits in every 14 may also be used [215].

4.1.3 DSRplus receiver

In the DSR receiver the DSR signal is converted to the quasi-DS1 level, the DSR decoder chip SAA 7500 from Philips Semiconductor normally being used (Fig. 4.3). With decoders of discrete design, such as the receiver from Technisat, the equivalent quasi-DS1 channels are also available in an unconcealed mode. The digital outputs of the SAA 7500 chip, a 16-bit parallel output and a serial I^2-S bus, are not used for DSRplus, since these output signals pass through a concealment circuit [211]. The DSR decoder chips have been designed with outputs, which provide a DSR corrected digital signal in a 63-bit BCH block and a DSR uncorrected in a 77-bit block. These outputs are marked DIC and DIE and carry the MUSICAM signals within a quasi-DS1 channel.

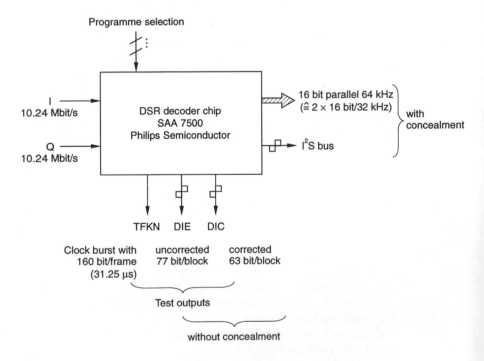

Fig. 4.3 DSR decoder chip SAA 7500 from Philips

With the aid of the clock signal TFKN, the demultiplexer circuit selects a MUSICAM signal which is decoded and returned to the DSR receiver to the point where the DSR signal would directly be taken from the DSR decoder. With the programme switch the receiver can either be set to the existing DSR or DSRplus (Fig. 4.4).

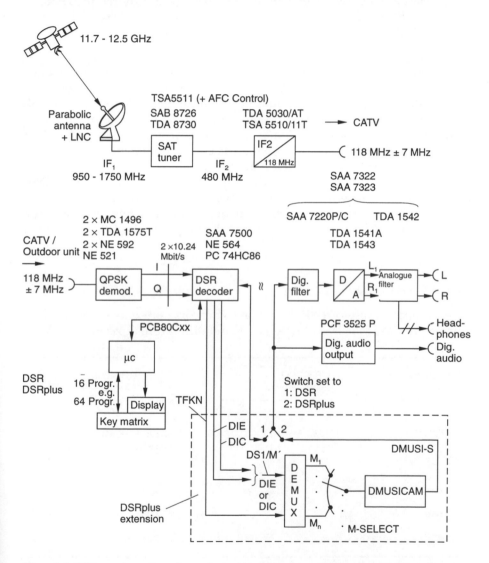

Fig. 4.4 DSR receiver (based on documentation from Philips-Semiconductor) with DSRplus extension

With DSRplus, the programme-associated data can be transmitted with the MUSICAM signal providing 2 Kbit/s per mono signal for auxiliary information. DS1 provides a free capacity of 20 Kbit/s for auxiliary information and in DSR a capacity of 32 Kbit/s is freely available within frame B.

Regarding the consumer receiver, an additional MUSICAM decoder chip (e.g. from Philips) with simple input selector logic is required for DSRplus (Fig. 4.4) [221].

The DSRplus solution has the following extra advantages:

- Employing MUSICAM coding with 96 Kbit/s per mono signal, three stereo signals plus one mono signal can be transmitted per DS1 channel with both DSR and MUSICAM protection and one stereo signal with MUSICAM protection only. Utilising a transponder with 16 DS1 channels in this way provides 48 stereo signals plus 16 mono signals with double protection and 16 stereo signals with single protection. For MUSICAM coding with 64 Kbit/s or 128 Kbit/s, the corresponding number of stereo signals can be seen from Table 4.1.
- DSRplus can use the existing DS1 line transmitters and the entire DSR satellite link including satellite and cable receivers, with the only necessary modification being data adaptation made on the quasi-DS1 channel at the beginning of the DS1 channel and at the output of the DSR decoder in the consumer's receiver.

4.1.4 Band limiting to 12 MHz

The compatible use of DSRplus in cable networks presents no problem as the bandwidth of the DSRplus signals do not differ from those of the ±7MHz

Transponder utilisation	L/R/MONO per DS1		L/R/MONO per transponder	
DSR	1/-		16/-	
DSR plus (with MUSICAM coding per MONO):	DSR and MUSICAM protected	MUSICAM protected	DSR and MUSICAM protected	MUSICAM protected
64 kbit/s	5/1	1/1	80/16	16/16
96 kbit/s	3/1	1/-	48/16	16/-
128 kbit/s	2/1	-/1	32/16	-/16

Table 4.1 Transponder utilisation for DSRplus (assuming integral number of L/R signals (stereo) per DS1 channel)

DSR. A possible strategy to introduce DSRplus in cable networks would be to open a second DSR channel in two special channels in Band III and to use them either completely, or initially only partly, for the DSRplus signals with the changeover to DSRplus implemented gradually. Another approach is to use the defined hyperband with 12 MHz channels, but for this use the DSR signal would have to be limited to a 12 MHz band.

The DSR signals have a frequency spectrum as shown in Fig. 4.5 and a 3 dB bandwidth computed to be between 13.3 MHz and 14.3 MHz. The energy components above the 14 MHz band limits may be suppressed at the expense of a slight increase of the carrier/noise ratio, assuming that the receiver spectrum shaping is dictated by the specification made by the remaining frequency range [216].

Depending on the relevant error rate, a reduction of the spectrum to a 12 MHz channel causes an increase of the carrier/noise ratio that cannot be neglected. Fig. 4.6 shows the spectrum limited to the 12 MHz bandwidth. The degradation of the error rate according to Fig. 4.7 yields a value of less than 1 dB for all relevant error rates, assuming spectrum shaping in the remaining signal in the receiver. Selection of the 12 MHz signal by a squarewave filter (corresponding to a D2-MAC receiver filter) means no further spectrum shaping, and yields a degradation of less than 1 dB for an error rate of 1×10^{-3} and less than 2 dB for 1×10^{-5}.

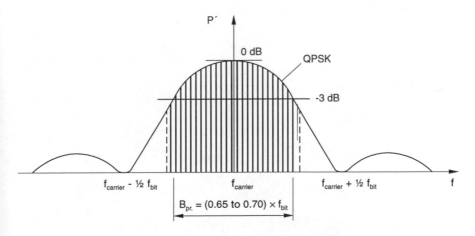

Fig. 4.5 Frequency spectrum of DSR/DSRplus signal

From this point of view the influence of the 12 MHz band limitation on the error rate is acceptable. The 3 dB bandwidth must be greater than or equal to 10.24 MHz to enable the receiver to reliably generate the clock rate, with 12 MHz DSR and hence 12 MHz DSRplus can be considered to be technically feasible [217].

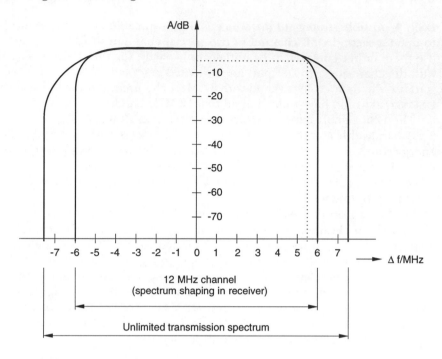

Fig. 4.6 DSR/DSRplus with 12 MHz band limiting

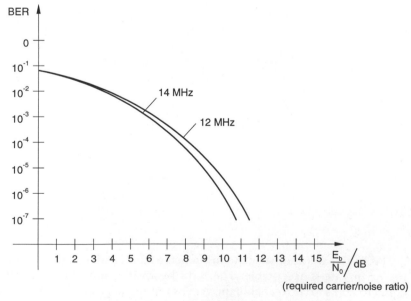

Fig. 4.7 Degradation as a function of band limiting to 12 MHz (spectrum shaping in transmission band of receiver) [216]

4.1.5 Field trial 'broadband communication network Munich'

During a field trial that was started in mid-1993 in Munich an additional local DSR packet was fed into the cable network. Part of the channels are used for measurement signals and a DSRplus packet. Signals are fed in at 356 MHz in the hyperband (centre frequency of channel H25; H25 results from the frequency range 302 MHz to 446 MHz are divided into the 12 MHz channels, H21 to H32, Section 1.3.2) or at 453 MHz with a bandwidth of 14 MHz and 12 MHz.

This field trial is to provide conclusive information on the vital questions on frequency compatibility, the feasibility of 12 MHz band limiting of the spectrum, bit error rates and adjacent channel compatibility [217]. The field trial phase is also planned to be extended to DSRplus test operation.

4.1.6 Introduction strategy

The DSRplus method has a capacity of 64 to 128 stereo signals, but will such an abundance of programmes be required? The list of programmes in analogue and digital format show that more than 100 programmes, mainly on TV subcarriers and in stereo, are currently broadcast via European satellites. This supports the fact that the programme variety already exists and far better economy could be acheived by concentrating transmissions on DSRplus, perhaps as a medium for programme distribution.

These 100 plus programmes are allocated on different satellites such as Intelsat, Eutelsat, DFS KOPERNIKUS, TV-SAT2, Astra, TDF1/2 and Télécom, with the satellite transmitter frequencies designated between 10.9 and 12.7 GHz. The modulation methods used include Wegener-Panda-1 (a compression method), DSR, MVR20, B-MAC, analogue and compressed FM and SAVE and TELL-Space. An example of 60 programmes contained in a DSRplus packet is given in [219].

DSRplus transmission in the 12 MHz channel provides a possibility of entering the American market with its 6 MHz channels. For the European DSR/DSRplus receiver industry this would create an enormous market expansion. In the US, several satellite radio systems are already commissioned. The DMX (digital music express) system offers 30 programmes in-house produced by DMX [218], 100,000 subscribers and a subscription fee of US$ 10. DMX plans to extend the service to offer up to 160 channels. Another service, Digital Cable Radio, started operations in May 1990 and currently boasts 2.4 million subscribers while Digital-Planet, another US system, followed in 1990 and has 400,000 subscribers.

DSRplus may be further developed to include access control and encryption (digital pay radio) on the Eurocrypt standard, an access control system for the MAC/packet family [214]. It features scrambling and access control at two levels.

DSRplus is the result of a strategic approach of introducing a combination of DSR and MUSICAM for 64 programmes on coaxial and/or fibre optic cable, for example, without affecting the introduction process of DSR. The iterative procedures that usually taking place upon introduction of a new service between specification, consumer equipment and the retail industry can be avoided with DSRplus. Data reduction with the aid of psychoacoustics by using MUSICAM gives enhanced performance features with minimum modification. DSRplus is suitable for stationary reception and for small satellite antenna systems with diameters of about 20 cm being suitable for portable reception.

In addition to DAB with its mobile reception qualities, DSRplus has its own opportunity in the future by offering a variety of programmes which a terrestrial system allowing for mobile reception could never provide.

4.1.7 Further MUSICAM applications

Table 4.2 shows the main audio coding methods used in the field of consumer equipment, sound broadcasting and TV sound transmission of the various TV systems. The coding methods using 16 bits/sample (CD), 16/14 floating point arithmetic (DS1), 14/11-bit companding (MSTD) or 14-bit linear coding (NICAM, D2MAC) are very transparent, i.e. the bit pattern/sample obtained in the receiver is congruent with that entered at the beginning of the transmission link. These audio channels allow a better use of psychoacoustic effects with practically no degradation of quality. This ensures that with the same channel capacity more audio programmes can be stored or transmitted, or with the same number of programmes the transmission bandwidth can be reduced.

The compact disc stores audio samples with 16-bit linear coding at a sampling frequency of 44.1 kHz. This means that the storage capacity for MUSICAM signals with 96 Kbit per mono signal is enhanced by a factor of seven.

A most interesting improvement and expansion of the Euroradio system can be achieved by suitable application of the DSRplus system. Euroradio can be considered as a transparent data transmission for DS1 with an interface in turn of 2×14 bit/32 kHz. Programme signals with reduced data, for example according to MUSICAM, can be inserted. This is particularly easy since the MUSICAM data rates are based on integral multiples of 32 kHz. By using for instance, MUSICAM with 96 Kbit/s per mono signal, a Euroradio system could be implemented for four stereo signals instead of just one. The same applies to Euroradio using the DSQ method (see Section 2.3.3.2).

The NICAM system may also be considered as a transparent data transmission system in which a data reduced source coding method, e.g. MUSICAM, can be used. This would give a greater sound capacity or with constant audio transmission capacity reduce the bandwidth of the

	Bits/sample	Sampling frequency kHz
CD	16	44.1
AES/EBU	16 to 24	48
DS1	16/14	32
DSQ	20 -> 18	48
MStD (Siemens, line transmission	14 -> 11	32
NICAM	14 -> 10	32
D2MAC	14	32
MUSICAM (96 kbit/s mono)	2	48
ASPEC (64 kbit/s mono)	1.3	48

Table 4.2 Main audio coding methods

transmission channel and so eliminate the out-of-band components in both the 7 MHz VHF and 8 MHz UHF channels.

4.2 Digital TV signals in introduced systems

4.2.1 Digital TV in DSR channel

This method applies to the transmission of TV signals via broadband media such as satellite, CATV, microwave or fibre optic links. The system combines two lines of development: the DSR method and digital video coding [223, 224].

An essential feature of DSR is the use of a transponder for the transmission of 16 stereo signals for stationary reception. At the transmitter end the signals are multiplexed on two datastreams each being 10.24 Mbit/s. The two data signals, I1 and Q1 in Fig. 4.8, are taken to a QPSK modulator which supplies the satellite IF signal at 70 MHz or 118 MHz ±7 MHz. On the DSR multiplexer/modulator in the earth station the I/Q interface is accessible via two plug-in links A and B.

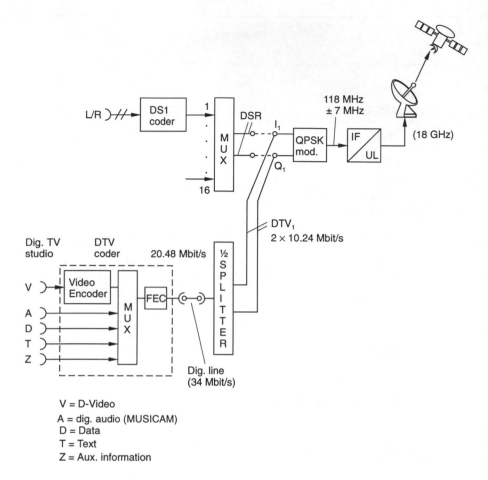

Fig. 4.8 DTV system at the generator end

In the standard DSR system, the cable headends are fed via the 18 GHz uplink and 12 GHz downlink and distribute the DSR signal at 118 MHz ±7 MHz in the special channels S2 and S3, to the CATV network (Fig. 4.9). For the DTV system, S9/S10, S19/S20 and the hyperband with 12 MHz bandwidth (Section 4.1.4) may be used for alternative frequencies. Consumer receivers, made by Philips, Grundig, Telefunken or Technisat, and are able to receive the DSR signal via cable or directly via a satellite receiving equipment with a dish or flat antenna, retailing from under DM1000 (£400).

Consumer DSR receivers use commercial LSI chips for QPSK demodulation and DSR decoding or discrete components. In the receiver,

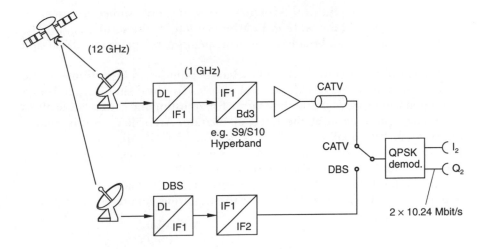

Fig. 4.9 DTV system via CATV and DBS

the interface between QPSK demodulator and DSR decoder is the 2 × 10.24 Mbit/s interface. A clock signal and a sync. signal are also available at this interface. Also available is a transparent 20.48 Mbit/s transmission system with the consumer receiver as an integral part in which the input bit sequence is equal to the output bit sequence (Fig. 4.10).

A quality digital TV signal close to HDTV and with a freely selectable baseband or channel coding can be fed into this 20 Mbit/s data transmission system. The frame structure of the data transmitted can be freely selected, to be similar to the DSR frame.

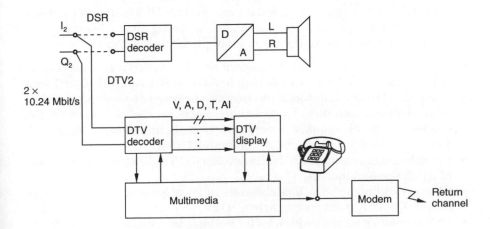

Fig. 4.10 DTV system at the receiver end

This method uses the process of picture coding and assumes that an error protection of 3 to 5 Mbit/s, 15 to 17 Mbit/s is left for the source signal. The state of the art allows broadcasting at a quality level between EDTV and HDTV.

Based on linear coding, video baseband coding uses algorithms for the bit rate reduction, uses psycho-optical effects. To put it simply, the bit rate reduction can be seen as a 'bit rate knob' for adjusting the coding algorithms to ensure optimisation at the selected bit rate. According to this model, optimum picture quality can be adjusted from 15 to 17 Mbit/s.

Instead of one high definition TV signal, it is of course possible to transmit four TV signals of standard quality [222].

By US standards, 20 Mbit/s is sufficient for HDTV quality:
The DigiCipher system [194, 197] has the following data rates:

Video (HDTV)	14.38 Mbit/s
Audio	1.76 Mbit/s
Data + text	126 Kbit/s
Control	126 Kbit/s
Total	16.4 Mbit/s
FEC rate	130/145
Total data rate	**19.43 Mbit/s**

The DSC-HDTV (digital spectrum-compatible HDTV) system has the following data referred to PAL/SECAM [199]:

Lines per picture	937/938
Pictures/s	50
Aspect ratio	16:9
Interlace	1:1
Line frequency	46.875 kHz

The DSC-HDTV system uses a data rate of 21.5 Mbit/s including video coder, data formatter and error control, i.e. including channel coding.

A 20 Mbit/s data transmission can be implemented as follows:

- The digital TV studio supplies component video signals with 4:2:2 coding parameters in line with CCIR Recommendation 601 and with a total data rate of 216 Mbit/s. Considerations may also be based on other source data rates defined according to a supranational multiplex hierarchy for gatewaying to digital postal networks (fourth hierarchical level with 140 Mbit/s or third hierarchical level with 34 Mbit/s).
- Video baseband coding with data reduction to 15 to 17 Mbit/s on the basis of psycho-optical models.
- Channel coding with the total data rate of 20.48 Mbit/s.
- Digital TV/DSR transmission system (DTV/DSR).
- DSR consumer receiver with 2 × 10.24-Mbit/s interface.
- Transition to digital TV receiver with microprocessor control, using, for

example, the IC set DIGIT 2000/3000 developed by ITT, which uses an internal 7 bit bus and a 17.7 MHz clock rate corresponding to 123.9 Mbit/s [164].

The 20.48 Mbit/s transmission can be configured using available components and adapting the data interface accordingly. The transmission channel may be a transponder following the DSR concept, or a transmission band in the cable network with two neighbouring TV channels of 7 MHz bandwidth each. A hyperband channel of 12 MHz bandwidth may be used as an alternative (Section 4.1.4).

Modern DSR receivers feature a continuously tunable frontend, enabling them to be tuned to several DSR channels. The DSR receiver can be used as a combinational unit for DSR, DSRplus and DTV according to the DTV/DSR method. Changing to the respective system is by switching on the 2×10. 24 Mbit/s interface (Fig. 4.10).

The DTV-DSR is suitable for stationary cable, satellite reception and quasi-portable reception provided that a 'visual contact' with the satellite is present. Taking the high power satellite TV-SAT as a basis, a signal in quasi HDTV quality or several TV signals of a standard quality can be received by dishes or flat antennas with a diameter of less than 30 cm.

The digital TV channel, which in addition to the video signal contains several sound channels — for instance based on the MUSICAM method — with programme-related additional information, may also carry an encryption which is activated within the receiver by a Pay TV or Smart Card descrambler .

The basic approach to DSRplus and DTV/DSR is intended for stationary and portable reception of 60 plus stereo programmes or one high definition TV programme or several standard definition TV programmes. For a larger TV capacity with improved quality or for HDTV transmission, bit rates of 30 to 40 Mbit/s are required. These data rates can be achieved by doubling the data capacity according to the DSR method. (Basic approaches are described in Section 4.3.)

4.2.2 Digital TV in conjunction with advanced audio recorder technology

In conjunction with digital TV via cable and satellite, multimedia and the recent developments in the field of audio recorders, this section deals with optical storage systems such as the video disc which has not yet made the breakthrough in consumer electronics.

The optical disc family includes the laser video disc, which is a nonerasable storage medium for professional video tapes. The analogue video signals are recorded on concentric spiral tracks running from the outside to the inside of the disc.

The pits on the digital video disc which are impressed by a highly focused laser beam give minimal reflection in a highly reflective environment. The digital video disc is available in three different versions: the read only, the write once/read many, and the erasable.

The OROM disc (optical read only memory) has a capacity of 500 Mbytes on one side.

The WORM video disc (write once/read many) has a storage capacity of 1.5 Gbytes per side.

The erasable laser optical disc in 5.25" format has a storage capacity of 400 Mbytes [242].

The development of these optical storage systems started in 1985. A comparison with the storage capacity of the more recently developed audio recorders DAT, MOD, DCC and MD shows that they are equal or superior to the video discs (Table 4.3). Assuming the video coding with SDTV quality (4.5 Mbit/s) and LDTV quality (1.5 Mbit/s), playing times could be obtained which are suitable for video and data source applications such as illustrated encyclopedias, video recorders or the recorder section of video cameras.

Storage medium	Gross channel capacity	Playing time with 1/2 rate channel coding for	
		SDTV (4.5 Mbit/s)	LDTV (1.5 MBit/s)
CD	15 Gbit	30 min	1.5 hrs
DAT	13 Gbit	25 min	1.15 hrs
MOD	3.5 Gbit	7 min	20 min
DCC	4 Gbit	8 min	23 min
MD	2.5 Gbit	5 min	14 min

Table 4.3 Playing times of audio consumer products used as video sources or video recorders

4.3 Optimised data capacity via satellite and cable

In Sections 4.1 and 4.2.1 methods have been described in which the use of the DSR channel has been improved by a factor of four due to MUSICAM coding or other kinds of information can be transmitted in the form of digital TV signals. All these methods ensure compatibility with the existing DSR system for consumer receivers.

This section now deals with transparent data transmission via satellite and cable, based on the DSR system. Consideration is given to noncompatible development by expanding the transmission bandwidth for the satellite medium or by increasing the modulation depth for satellite and cable to enhance the transmission capacity for further applications.

4.3.1 Two DSR channels per transponder

The first approach is to transmit two DSR channels within the bandwidth of a transponder. In the DSR system, the DSR signal is provided with approximately 14 MHz bandwidth via the 27 MHz transponder. Band limiting to 12 MHz is possible but at the expense of degradation of less than 1 dB of the carrier/noise ratio (Section 4.1.4). Therefore, two DSR channels can be transmitted on one transponder featuring a bandwidth of at least 27 MHz (Fig. 4.11). Within the transponder bandwidth, the two DSR channels may be adjacent or in an upward compatible solution with the second DSR channel possibly split into two halves taking the free space available at the top and bottom of the band. With this approach, the data transmission rate on the transponder is doubled to give 40 Mbit/s [226]. With reduced transponder power the error rate of the DSR transmission is measured and the terrestrial coverage (footprint) reduced.

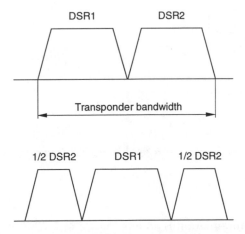

Fig. 4.11 Arrangement of two DSR channels per transponder (bandwidths of e.g. 27 MHz, 32 MHz (Astra), 36 MHz (KOPERNIKUS), 72 MHz (20/30 GHz KOPERNIKUS)

4.3.2 8PSK/2AM modulation

The second approach to increase the DSR data rates uses a modulation depth greater than that of the QPSK modulation.

For satellite broadcasting, a modulation mode with constant carrier amplitude, e.g. FM or QPSK modulation, is predominantly used. The next step would be 8PSK or 16PSK modulation. However, 8PSK has the disadvantage of the resulting data rate being increased by a factor of only 1.5. 16PSK gives double the data rate, but the required C/N is about 3.5 dB higher compared with 16QAM.

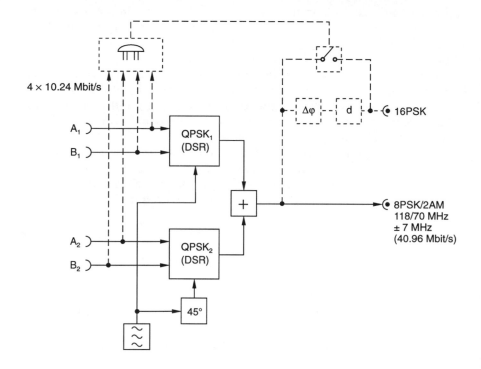

Fig. 4.12 8PSK/2AM modulation, here for double DSR data rate; dashed: conversion into 16PSK

The 8PSK/2AM modulation is an alternative to 16PSK and 16QAM and is obtained by simple vector addition of the output signals from two QPSK modulators according to the DSR principle (Fig. 4.12) [227]. The signal at the output of the adder has eight phase positions and two amplitude levels, corresponding to the amplitude ratio of the two diagonals of an equilateral parallelogram with the angles of 45° and 135° (Fig. 4.13). The vector diagram and the time function of 8PSK/2AM are shown in Fig. 4.14 and Fig. 4.15. The inner modulation circle is attenuated by 7.66 dB against the outer circle, corresponding to 20log (tan 22.5°).

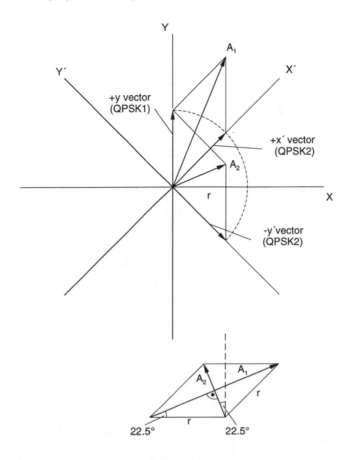

Fig. 4.13 Vector diagram derived for 8PSK/2AM

Note:

- The level relationship of the two modulation circles can be varied by a switched attenuator for the modulation points on the outer circle.
- With an input logic, 8PSK/2AM can be turned into pure 8PSK, so only the modulation points on the outer circle will be addressed (half data capacity).
- 8PSK/2AM can be converted into pure 16PSK by inserting a switchable 22.5° phase shifter and 7.66 dB attenuator at the signal output. The phase shifter and the attenuator only becomes effective for the outer modulation points (dashed lines in Fig. 4.12).

As illustrated in Fig. 4.14, the decision circles of the modulation points on the outer circle differ in size from those on the inner circle. This can be compensated in data transmission by channel coding providing the

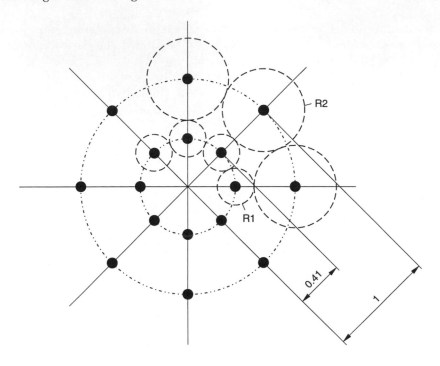

Fig. 4.14 Vector diagram with decision circles for 8PSK/2AM

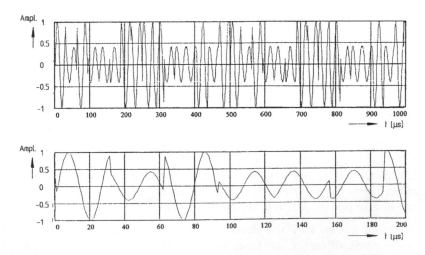

Fig. 4.15 Time function of 8PSK/2AM signal (bottom: expanded) [227]

modulation points on the inner circle with a higher error protection from those on the outer circle. If, for instance, the FEC is 3/4, the error protection capacity is 5 Mbit/s for a net data transmission capacity of 15 Mbit/s. Two thirds of the error protection capacity, 3.3 Mbit/s, could be used for the inner circle and one third, 1.7 Mbit/s, for the outer circle. This (in terms of the modulation points) unequal error protection requires a small number of organisation bits to handle the two classes of modulation points.

For lab and field trials, 8PSK/2AM can simply be implemented by two coder/modulator series modules, a 45° phase shifter for the carrier frequency, and a simple adder stage for the vectors of the two vector diagrams rotated by 45°. For professional operation, a solution with optimised circuitry should be sought [226].

4.3.3 Comparing 8PSK/2AM and 16QAM

With 16QAM (and similarly with 8PSK/2AM) four bits are transmitted per modulation step, giving 16 modulation points. Fig. 4.16 shows the block

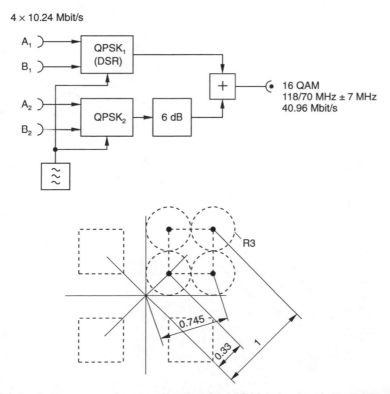

Fig. 4.16 Block diagram and vector diagram for 16QAM, for double the DSR data rate

	Amplitude states	Phase states	Modulation points	C/N in dB for $p_e = 10^{-6}$	
2PSK	1	2	2	10.7	
4PSK	1	4	4	13.7	
8PSK	1	8	8	18.8	
16PSK	1	16	16	24.0	
16QAM	3	12	16	20.5	
32QAM	5	28	32	24.0	
64QAM	9	52	64	27.0	
8PSK/2AM	2	8	16	18.9	outer 8PSK
p_e = mean bit error probability				26.5	inner 8PSK

Table 4.4 Digital modulation methods in C/N comparison [220, 225]

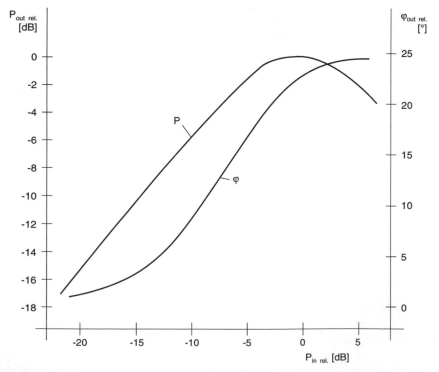

Fig. 4.17 Power and phase characteristic of a TWTA

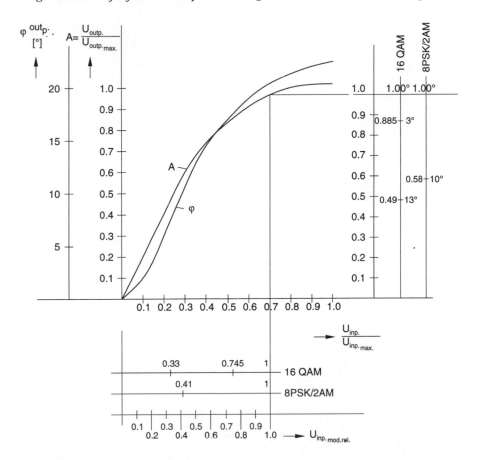

Fig. 4.18 16QAM and 8PSK/2AM via transponder TWTA at half the maximum input power

diagram of 16QAM and its easy implementation. Two QPSK modulators are required, with the output signal of one of the modulators being attenuated by 6 dB. The vector diagram shown at the bottom of Fig. 4.16 is obtained via an adder and the modulation points being equidistant between neighbouring quadrants.

For comparing 8PSK/2AM and 16QAM, at least three criteria have to be considered:

- required C/N
- echo susceptibility and
- driving of the power amplifier stages.

Regarding the required C/N, a comparison of 8PSK/2AM with 16QAM yields a slight advantage for 16QAM [225]:

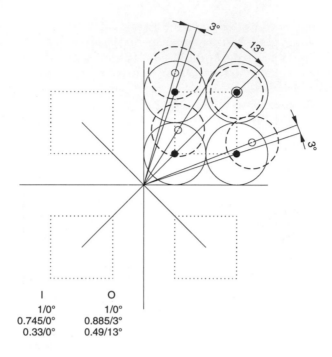

I   O
1/0°   1/0°
0.745/0° 0.885/3°
0.33/0°   0.49/13°

Fig. 4.19 16QAM following transponder with TWTA: I = input amplitude and phase;
O = output amplitude and phase error

16QAM =	20.5 dB
8PSK/2AM =	21.2 dB
	18.9 dB outer modulation points
	26.5 dB inner modulation points

Note: 8PSK/2AM may of course also be considered as a pure 8PSK with improved C/N compared to 16QAM and an additional data channel of the same capacity with approximately 7.5 dB lower C/N.

The QAM method has a drawback regarding echo susceptibility, since there are 12 phase states and 3 amplitude states compared to 8 phase states and 2 amplitude states in the 8PSK/2AM mode (Table 4.4).

Another criterion for comparing these two modulation modes is the nonlinear transmission characteristic of the TWTA (travelling wave tube amplifier) in the transponder. Fig. 4.17 shows that with 16QAM the TWTA cannot be fully driven. Even when driving at half the maximum input power, the transmission characteristic is nonlinear (see Fig. 4.18).

With 8PSK/2AM modulation, a better transmission characteristic according to the magnitude and phase is obtained, as shown in Fig. 4.18 [228].

The nonlinear transmission characteristic can be compensated by taking appropriate measures. Precorrection in the satellite transmitter, assuming conditions are stable, is one possibility. In the case of drift in the transponder nonlinearity, automatic precorrection control via a satellite receiver can be employed. It is, however, more advantageous to use a modulation mode with less amplitude and phase steps in the nonlinear transmission channel. The decision circles for transmission without precorrection are shown in Fig. 4.19 and Fig. 4.20.

4.3.4 32QAM and 64QAM

The data capacity can be further enhanced by adopting 32QAM or 64QAM. 64QAM can be implemented on the DSR principle by connecting in parallel three modulators with an amplitude ratio of 1:0.5:0.25. In the vector diagram shown in Fig. 4.21, 16 modulation points are contained in each quadrant.

32QAM can be derived from 64QAM by way of an input logic, producing from the 5-bit input pattern a 6-bit output pattern. By this method, 32

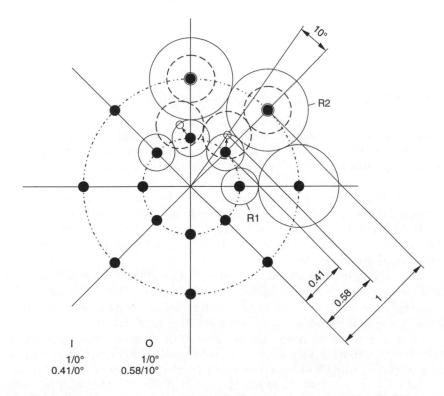

	I	O
	1/0°	1/0°
	0.41/0°	0.58/10°

Fig. 4.20 8PSK/2AM following transponder with TWTA: I = input amplitude and phase; O = output amplitude and phase error

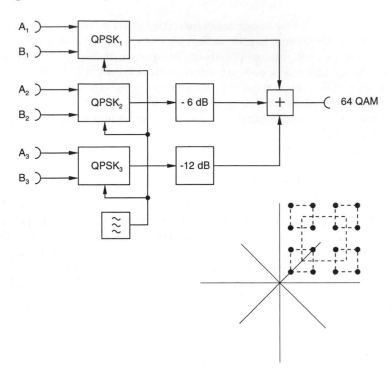

Fig. 4.21 Block diagram and vector diagram of 64QAM

modulation points are addressed and 5 bits per step transmitted. The advantage of 32QAM compared to 64QAM is that the radius of the decision circles is greater by a factor of $\sqrt{2}$, accounting for a higher reliability of the transmission channel.

Generally, a high-order modulator can be configured from QPSK elements, the amplitude for the single modulator being freely selectable by inserting an attenuator and the phase determined by inserting a phase shifter. The total data capacity can be reduced by an input logic for a selection of the modulation points.

With a four-level modulation mode (8PSK/2AM or 16QAM) and two DSR packets, 80 Mbit/s can be reliably transmitted. This data rate can be achieved for instance by adding simple functional elements such as phase shifters, attenuators, adders or filters to existing modulators.

Regarding the data transmission capacity of a cable it can be assumed that using a 16-level modulation technique 40 Mbit/s per step can be transmitted in two TV channels or one hyperband channel. In view of the typical C/N ratio of ≥40 dB of the cable, 32QAM or 64QAM may be used. With 64QAM, a data rate of over 30 Mbit/s per TV channel (7 MHz) and over 60 Mbit/s in a hyperband channel (12 MHz) can be attained.

Irrespective of the modulation mode, trellis-coded modulation and Viterbi decoding can be adopted for channel coding and decoding to yield a significant modulation gain, i.e. enhanced transmission reliability but with the useful data rate reduced (Section 4.7.2).

4.4 Programme distribution to sound broadcast transmitters

4.4.1 Distribution to VHF FM transmitters

Using the conventional method, programmes are distributed to the VHF FM transmitters via wired line networks and microwave links but the great disadvantage of this method is the high rental fees. Alternatively, programme distribution can be provided via relay reception with several slave transmitters receiving the signal from a master transmitter or from a slave transmitter with multiple cascading. Due to the increasing number of programme providers, the frequency-intensive relay receiving technique can be subjected to losses in quality, and therefore programme distribution to FM transmitters via satellite using digital methods, e.g. DSR, are more

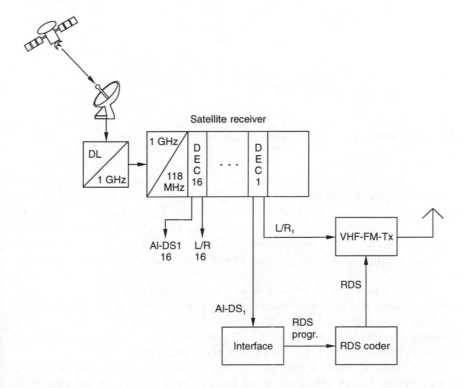

Fig. 4.22 Programme and RDS distribution to the VHF FM transmitter using DSR

frequently considered (Fig. 4.22). Professional DSR satellite receivers have been developed to provide the analogue L/R signal and — at a separate interface — the auxiliary information contained in DS1 [231].

The professional DSR receiver consists of two basic modules: the digital sound convertor and the digital sound receiver.

In the convertor the QPSK modulated signal is converted from the first satellite IF of 1 GHz into 118 MHz. The operating output, which is muted in the case of a failure, has an additional convertor fitted with a permanently through-connected test output while integral tuning aids adjust level alignment and frequency [230].

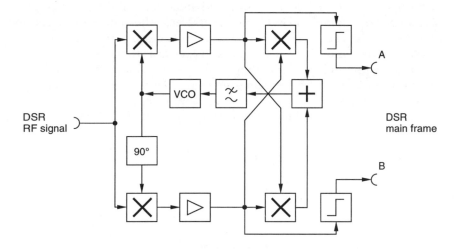

Fig. 4.23 QPSK demodulator in DSR receiver with Costas loop

The subsequent digital sound receiver processes the QPSK modulated 118 MHz signal. For demodulating the QPSK signal, the reference phase has to be retrieved from the received signal. As the QPSK spectrum does not contain any carrier components, the reference phase has to be derived from the QPSK signal. This is achieved by the Costas Loop, a special phase control loop for regulating a VCO (voltage controlled oscillator) so the reference phase of the transmitter signal is regenerated. The two datastreams A and B can be reconstructed by means of a sync. demodulator (Fig. 4.23). After demodulation and decoding the selected programme is taken via a fourfold oversampling digital filter to the 16-bit D/A convertor. The subsequent steep-skirt audio lowpass filter eliminates the aliasing components which may cause audible interference in the case of remodulation.

The DSR receiver is fitted with an audio substitution signal input, switchover to the substitution signal being controlled by an integral BER meter. The switchover threshold is set to a bit error rate of approximately

5×10^{-4}, since with DSR transmission audible errors occur only above a BER of about 3×10^{-3}. Moreover, the DSR receiver monitors the auxiliary information associated with the selected channel. If the message 'no programme' is sent, switchover to the substitution signal takes place.

The DSR receiver is suitable for programme distribution and for distributing the dynamic data in the RDS (radio data system). The programme-associated RDS data must be generated in the studio and for operational reasons taken to the RDS coder in the transmitter station similar to the programme signals.

DSRplus may also be used for programme distribution instead of the DSR method. The only modification needed to make this possible is to connect a demultiplexer logic circuit for the selected MUSICAM signal and a MUSICAM decoder to the quasi DS1 output of the DSR receiver. The programme-associated RDS data can be transmitted in the auxiliary information (AI) channel in DS1 or in MUSICAM mode, providing 2 Kbit/s per mono signal for auxiliary information.

4.4.2 Distribution to DAB transmitters

4.4.2.1 Analogue OFDM distribution
Programme distribution to DAB transmitter stations via the 30/20 GHz satellite KOPERNIKUS by a pseudo video signal has been described in Section 2.5.7.2. According to this method, four analogue OFDM packets are transmitted in FM mode in a 7 MHz baseband [236]. This solution has the advantage that in the field trials the DAB signal was distributed to the DAB transmitter therefore removing the need for coding or modulation in these transmitters. Moreover, the DAB requirement for bit-synchronous radiation by each transmitter in a single frequency network is already satisfied. The 20/30 GHz transponder was simply chosen because of its availability. The disadvantages of this method of programme distribution which become apparent during the field trials are listed below [208]:

- Frequency conversion of the pseudo video signal at the transmitter before it is sent to the satellite and filtered by the DAB transmitters is very complex.
- No bit error correction can be applied prior to terrestrial broadcasting.
- Six programmes per OFDM signal provide a total of only 24 programmes. A considerably greater number of programmes is required for national coverage.
- Transmission via the satellite yields a useful capacity of approximately 4×1.2 Mbit/s; by comparison, with DSR a useful capacity of approximately 15 Mbit/s is transmitted, with a further increase by a factor of 2 to 4 being possible (Section 4.3).
- Another disadvantage results from the effect of the delay difference

between a satellite signal received in the north of the single-frequency network and one received in the south (Section 4.4.2.2). The delay difference has to be compensated by the DAB transmitter station in avoiding a loss in the guard interval time. With an analogue signal feeding the DAB transmitters, an analogue delay element is required. An A/D and D/A digital method of conversion is chosen because of the required long delay times .

Further disadvantages result in conjunction with the 30/20 GHz system:

- A spare link is not available in the 30/20 GHz range.
- The 20 GHz compared to the 12 GHz receiving equipment is more complex. Dishes of 1.5 to 2 m in diameter and with high directivity are the most suitable, and it may be necesssary to heat and readjust them.
- The 20 GHz transmission corridor is so narrow that the swing of the satellite may critically affect the receiving field strength.
- The 30/20 GHz link is susceptive to adverse weather conditions such as rain, fog and hail resulting in signal fading of between 20 and 30 dB.

For DAB system operation, satellite programme distribution should therefore be in digital form and the satellite itself chosen as the medium because of the large number of DAB transmitters in the single-frequency network that have to be supplied. Since the transmitter signal comes from a single source, the satellite, the problem of bit-synchronous radiation required in the DAB network can therefore be solved quite simply. A disadvantage of the satellite transmission is that due to a chance of transmission failure some spare capacity must always be available. This may take the form of a standby transponder on the same satellite, or preferably another satellite. Another adverse effect is the delay of about 240 ms on the satellite link, which becomes particularly evident on live broadcasts with viewers' participation. A further disadvantage that has already been mentioned is the site-dependent delay of the satellite signal that has to be compensated by the DAB transmitter station.

4.4.2.2 Propagation time difference in satellite distribution

The tolerance of clock- and bit-synchronous radiation of the DAB transmitter signal in the single-frequency network is determined by the guard interval. In mode 1 of the DAB specification the guard interval is 250 µs. Allowing for a synchronisation error of 5%, the permissible time difference Δt is 12.5 µs. Using $c \times \Delta t$, this time difference can be converted into the corresponding transmission distance. The resulting maximum permissible difference in the downlink of two neighbouring transmitters is 3.75 km. With a distance of about 60 km between the transmitters, the downlink difference and the propagation time difference becomes correspondingly greater.

The length of the downlink to the individual DAB transmitters can be determined exactly by two- and three-dimensional trigonometry. To calculate the geostationary orbit of the satellite, the longitude and latitude of the transmitter site have to be known [229].

In the following example, the effect of a longitude and latitude variation of two DAB transmitter sites is to be investigated. To determine the effect of latitude variation, a transmitter chain is assumed to be positioned at 11°30'00" longitude, roughly corresponding to the longitude of Munich. Table 4.5 shows that even slight variation of the latitude by the order of a few minutes causes a greater difference than allowed. At the considered latitude of 48° (Munich) a variation of ten minutes corresponds to a distance of about 18.7 km. Two DAB transmitters at a distance of 30 km from each other would receive and radiate the DAB signal with a time difference of 75 µs, considerably reducing the guard interval.

The same problem is encountered for variations of the longitude, the latitude of Munich being kept constant at 48°10'00". According to Table 4.6, the differences in the propagation time are not as significant as with a variation of the latitude. The maximum synchronisation time of approximately 10 µs is also exceeded.

N	Longitude (degree, min, sec)	Transmitter distance to N-1 (km)	Downlink difference ref. to N-1 (km)	Time difference ref. to N-1 (µsek)
1	09 00 00	-	-	-
2	09 15 00	18.5	5.07	16.8
3	09 30 00	18.5	4.98	16.6
4	09 45 00	18.5	4.89	16.3
5	10 00 00	18.5	4.81	16.0
6	10 30 00	37.0	9.36	31.2
7	11 00 00	37.0	9.02	30.1
8	11 30 00	37.0	8.67	28.9
9	12 00 00	37.0	8.32	27.7
10	12 30 00	37.0	7.98	26.6
11	13 00 00	37.0	7.62	25.4

Table 4.5 Effects of longitude variation on satellite transmission [229]

For synchronous operation in the single-frequency network, a time delay element for delay equalisation is required in the DAB transmitters. The digital delay element can be implemented by a DSP chip or FIFO (first in/first out) memory.

N	Latitude (degree, min, sec)	Transmitter distance to N-1 (km)	Downlink difference ref. to N-1 (km)	Time difference ref. to N-1 (μsek)
1	48 00 00	-	-	-
2	48 10 00	18.7	14.87	49.6
3	48 20 00	18.7	14.90	49.7
4	48 30 00	18.7	14.93	49.8
5	48 40 00	18.7	14.97	49.9
6	48 50 00	18.7	14.99	50.0
7	49 00 00	18.7	15.03	50.1
8	49 30 00	56	45.28	150.9
9	50 00 00	56	45.56	151.9
10	50 30 00	56	45.84	152.8
11	51 00 00	56	46.12	153.7
12	51 30 00	56	46.39	154.6
13	52 00 00	56	46.66	155.5

Table 4.6 Effects of latitude variation on satellite transmission

For adjusting the time delay at each transmitter site, the downlink of the most adverse transmitter site in the reception area has to be determined. This downlink can be converted into the corresponding time t_{max} A time reserve has to be included to allow for the different response times of the different makes of transmitters.

Adjustment of the delay time for each transmitter site can be automated and the nominal value transmitted to the transmitters together with the programme via satellite. By the transmitter co-ordinates, the actual time delay at the transmitter station can be calculated and the difference from the nominal value adjusted. The nominal value can thus be easily varied, for example, upon a change of satellite.

4.4.2.3 Digital programme distribution on the basis of MUSICAM

For nationwide coverage of a country with a regional structure, four frequencies are for instance required for single-frequency networks. A larger country may have, for instance, ten regional networks with six programmes, totalling 60 programmes transmitted via the satellite. This problem can be solved by using the DSRplus method (Section 4.1). The basis of this method is DSR, with MUSICAM coding applied at the signal source and channel coding, multiplexing and OFDM modulation in the DAB transmitter (Fig. 4.24). One problem to solve is word- and clock-synchronous, i.e.

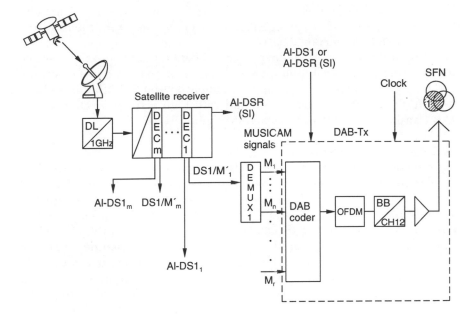

Fig. 4.24 Programme distribution to DAB transmitters using DSRplus method

bit-synchronous, radiation by the DAB transmitters in the single-frequency network. Clock synchronism is ensured by the DS1/DSR method. Word synchronism is implemented via the available auxiliary information channels in the DS1 or DSR frame. A net capacity of 24 Kbit/s is available per DS1 channel, while the DSR main frame B has a free data capacity of 32 Kbit/s [230].

Note: Clock and frame synchronisation of the DAB transmitters can also be implemented by GPS, which, additionally to navigational data, can supply high-precision time information. The GPS receiver with the navigational computer can determine from the GPS satellite signals the pseudo distances to use this information in setting the time on the receiver clock. From the inaccuracy of the position measurement by 35 m to 100 m for the C/A code (civil application) and 20 m to 30 m for the P code (military application) a time standard accurate to within microseconds can be derived with the aid of the light velocity. This time standard allows sufficiently accurate clock synchronisation of the DAB symbols in all three modes (guard intervals 250 µs, 62.5 µs and 31.25 µs). Frame synchronisation (frame duration in modes 1, 2 and 3: 96 ms, 24 ms and 24 ms) is achieved by an absolute time protocol.

If programmes are distributed to the DAB transmitters by the MUSICAM method, external data such as service information or service configuration data have to be transmitted, from which the fast information channel and the synchronisation channel signals are taken by the DAB coder in the transmitter station (Section 2.5.2).

These additional external data require a data rate of approximately 100 Kbit/s in the DAB signal multiplex and the fixed code rate of 1/3 on the feed link, which corresponds to approximately 35 Kbit/s. With the DSRplus method, these data may be transmitted with high error protection to the DAB transmitter instead of a MUSICAM signal [91, 92].

4.4.2.4 DAB/VHF-FM simulcast

It is intended to operate DAB and VHF/FM simultaneously for more than 15 years. According to Fig. 4.25, it is only necessary to decode the MUSICAM signals at the input of the DAB transmitter, then to convert from digital to analogue before applying them to the VHF/FM transmitters. The problem of programme-associated RDS data is solved by transmitting the MUSICAM programme-associated data in the auxiliary information channel of DS1 in an addressed form while the MUSICAM signals are transmitted in the DS1 channel. The DSR decoder in the DAB transmitter station supplies this auxiliary information in separate channels. After demultiplexing, the programme-associated RDS data are then applied to the relevant RDS coders.

An alternative is to transmit the RDS data in the MUSICAM auxiliary information channel having a capacity of 2 Kbit/s for programme-associated data.

4.4.2.5 Programme distribution on the basis of channel-coded DAB multiplex signals

This approach is based on channel-coded DAB signals of 2.4 Mbit/s for satellite transmission (Fig. 4.26). Eight such digital OFDM packets (corresponding to $8 \times 6 = 48$ programmes) are, for example, transmitted via the DSR channel through a processor-controlled multiplexer. The remaining data capacity resulting from the difference between the gross data rate of 20.48 MHz and 8×2.4 Mbit/s = 19.2 Mbit/s is used for an adapted data frame or for auxiliary information [226].

The requirement for bit-synchronous radiation is fully met since the DAB transmitter is only followed by the OFDM modulator, frequency convertor and linear amplifier.

For programme distribution by DSR, the methods described in Section 4.3 for an increase in the data rate (greater modulation depth, and the use of two DSR channels on one transponder) can also be used.

The DAB multiplex can be generated in the earth station. In this case a transport multiplex of 2 Mbit/s is used in the DS2 line network prior to satellite transmission.

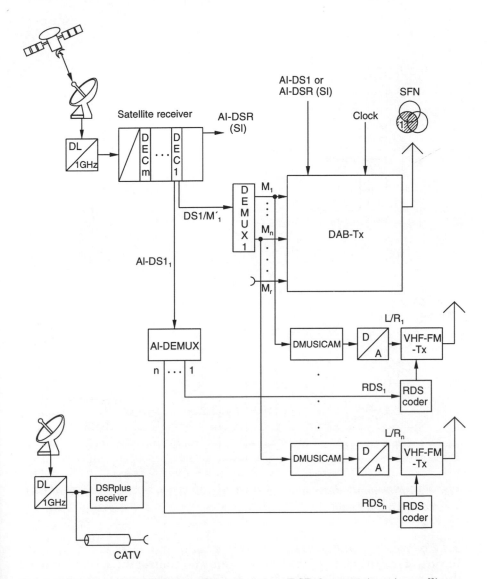

Fig. 4.25 DAB and VHF-FM simulcast operation (DSRplus reception via satellite or CATV possible)

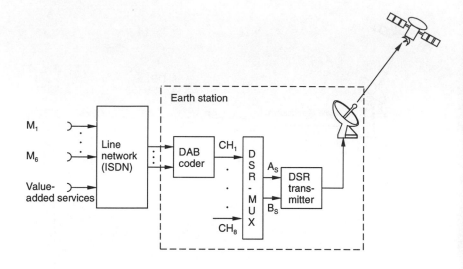

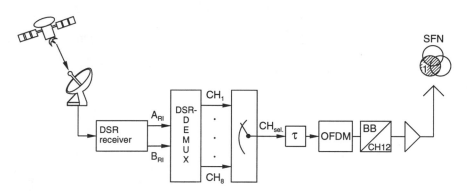

Fig. 4.26 Programme distribution to DAB transmitters (e.g.) on the basis of DAB-channel coded signals

4.4.2.6 Programme distribution on the basis of SCPC

In view of the structure of the broadcasters who control the service between the programme sources and listener, feeding the programme to a common ground station for transmission to the satellite may sometimes be a disadvantage, regarding both the rental fees for the programme lines (e.g. ISDN, DS1, DS2) and the functional and operational aspects.

The SCPC (single channel per carrier) method is an alternative in which one satellite transponder is driven by several earth stations. The satellite transmitter signal is the sum of the signals received. Using SCPC, a narrowband QPSK system (NB-QPSK) can be configured. As shown in Fig. 4.27, a DAB signal is constructed first with MUSICAM, channel coder and

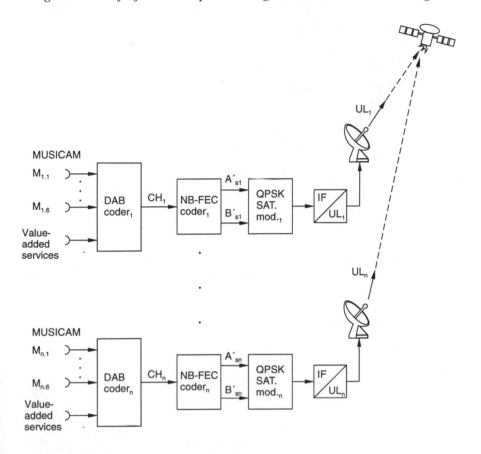

Fig. 4.27 SCPC method on the basis of DAB channel-coded signals (transmitter end)

multiplexer. With the aid of an additional frame in the NB-FEC block, the DAB-CH signal with 2.4 Mbit/s is then coded to form the signals A_{s1}' and A_{s2}' with, for example, 2×1.5 Mbit/s with the QPSK modulator next in the process. The transmitter signal has a bandwidth of 2 MHz (Fig. 4.27). A number of n such earth stations feed the satellite, giving a transponder structure as shown in Fig. 4.28. In the DAB transmitter station, an NB-QPSK block is filtered, QPSK demodulated and applied to the DAB transmitter amplifier via the OFDM modulator (Fig. 4.29) [226].

With $n = 4$, 24 programmes in total can be transmitted via one transponder ($n = 6$: 36 programmes). The interfaces A'/B' can be considered to be transparent, so that transmission is also possible on the basis of other data parameters (special DAB transport multiplex with 2.048 Mbit/s (DS2)) or other programme numbers (1 to 5).

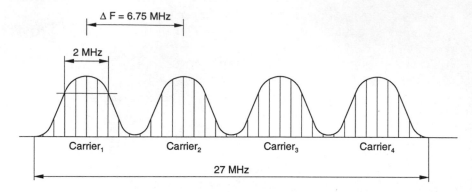

Fig. 4.28 SCPC signals on transponder (n = 4)

With this method, the requirements for scrambling, e.g. on a MUSICAM basis or access control, e.g. in line with the EUROCRYPT standard, can easily be met via the NB-FEC block. The narrowband method can also be combined with other services, such as TV signals, on a transponder but only if there are appropriate gaps of, say, 1 or 2 MHz.

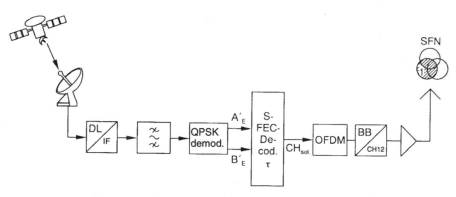

Fig. 4.29 DAB transmitter with SCPC reception

The SCPC sum signal on the transponder, in which four (or *n*) QPSK modulated signals with constant amplitudes are added up, is comparable to a quasi white noise signal, since within the given bandwidth almost all frequencies occur, although with different amplitudes. Driving the transponder with this signal featuring a high crest factor has yet to be tested, since the TWTA of the transponder exhibits a strong nonlinear characteristic (Section 4.3).

Programme distribution to the DAB transmitters on the basis of a digital DAB interface could lead to the development of a combination of DAB receivers for the mobile reception in terrestrial networks and for the stationary reception from the satellite transponder of high programme capacity.

4.5 Model for narrowband DAB

The DAB method according to EUREKA 147 is based on an OFDM packet, in which, for instance, six stereo programmes are radiated by a DAB transmitter. This does not correspond to the present universal practice of one programme per sound broadcast transmitter. In particular in the US, in-band solutions with digital programme transmission within the given bandwidth of the present sound broadcast structure will increasingly become adopted.

There is also a proposal from Bosch/Blaupunkt based on the modulation scheme in which each programme is separately coded and radiated, not with one constant carrier, but several carriers changing in small time intervals. In that procedure some programmes may be transmitted in a specific bandwidth where each programme carrier is changing in time slots (frequency hopping). The hopping organisation is predefined with the receiver tuned synchronously to the transmitted frequency.

With this procedure the disadvantage of the narrowband concept with a constant carrier, namely that the received signal might be eliminated by Rayleigh fading, can be evaded by changing the frequency within the spectrum provided for DAB — e.g. *part* of a TV channel — according to a fixed scheme.

This works as follows. If reception at a specific narrowband frequency is critical, this can be compensated by the transmission channels at the preceding and the subsequent hopping frequency. In conjunction with time interleaving a result can be achieved similar to a broadband method. Disadvantages of the hopping method are that high-speed oscillators have to be used in the receiver and that the hopping organisation requires part of the transmission capacity [86].

In the following alternative, a true narrowband transmission is implemented with an enhanced data transmission capacity allowing all known measures for error correction and modulation gain to be employed while solving the fading problem by antenna diversity at the receiver.

Based on the EUREKA-147 system, mode 1, 200 carriers could, for instance, be used for a bandwidth of 200 kHz, giving a gross data rate of about 300 Kbit/s when using QPSK modulation and by also taking into account the guard interval. With MUSICAM at 96 Kbit/s mono and some data transmission capacity being used for value-added services, a channel coding rate of less than 2/3 would result. Instead of QPSK modulation, a

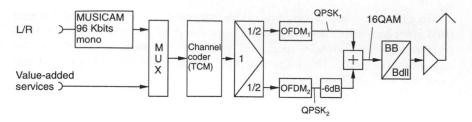

Fig. 4.30 Example of narrowband DAB transmitter with 16QAM

16-level modulation mode, e.g. 8PSK/2AM or 16QAM (Section 4.3) may be used. This solution is based on two QPSK-OFDM modulators, which in the case of 8PSK/2AM operate with a co-ordinate system rotated by 45°, and in the case of 16QAM with vectors that differ by 6 dB, with the output signals combined by an adder (Fig. 4.30). With the gross data rate of approximately 600 Kbit/s gained in this example, all measures of digital transmission can be employed: guard interval, time and frequency interleaving, channel coding rate $\geq 1/2$, trellis-coded modulation, and graceful degradation.

To overcome the problem of Rayleigh fading in VHF FM sound broadcasting, space diversity or antenna diversity systems can be used, which uses two to four staggered omnidirectional or directional antennas. To ensure optimum functioning of the antenna system, the antennas must have a minimum spacing of about half the wavelength $\lambda/2 = 1.5$ m. Through a switching system the antenna which offers the best reception quality is selected, i.e. only one antenna in use is contributing to the useful signal while the others are inactive. In a configuration with one tuner all antennas have to be scanned and the values of the associated reception quality stored in a buffer. The adverse effects of this concept are the interference caused by scanning and the problem of adaptation in the case of rapidly changing receiving conditions while in mobile operation [232, 233, 234].

A significant improvement can be achieved by using an adaptive antenna system, in which — in contrast to the switching diversity — all antennas in use are switched to the receiver via amplitude and phase control elements. The ADA (auto-directional antenna) system developed by Blaupunkt utilises the signals from the individual antennas after conversion into the IF by modifying them in amplitude and phase and combining them to yield an optimal sum signal. The separate FM tuners and IF mixers allocated to each antenna have a common oscillator for distortion-free conversion of amplitudes and phases into the IF. The modules following the sum signal adder stage correspond to those of a conventional VHF receiver. The maximum sum power can be used as a criterion for the amplitude and phase adjustments by the control elements. For this purpose the phase difference between the sum signal and the individual output signals of the tuners is measured, from which the optimum settings of the control elements are derived through a processor unit.

The response of the ADA system has been simulated with four isotropic radiators. Under multipath receiving conditions with delay differences between 10 μs and 80 μs, stationary directional antenna patterns were obtained in about 2 ms. This settling period ensures optimum system adaptation in mobile operation.

The single-tuner Scandiv system (scanning antenna diversity system) for VHF-FM developed by Fuba [235] is an alternative to the antenna diversity system but with a tuner allocated to each antenna. The short fault detection time of 25 μs allows sequential testing of several antennas without causing any acoustic disturbance. In the Scandiv system, the 10.7 MHz IF signal is FM demodulated; if the audio signal is subjected to interference caused by multipath reception, noise or too low a level, a switch to the next antenna is made.

It is not a major problem to install two to four antennas on one vehicle. One main antenna is provided for VHF and for LF and MF. The other antennas are for VHF and are designed to be either as a rod or a windscreen antenna. Antenna elements have recently been constructed to be integral in the hollow space of plastic bumpers.

The antenna diversity systems designed for VHF FM sound broadcasting can also be used for narrowband DAB (Fig. 4.31). For the use of antenna diversity in narrowband DAB systems it is of key importance to achieve the required minimum field strength for an S/N of more than 10 dB. The phase rotation of the signals in a quadrant can be used as a criterion of the antenna control elements. The signals in the quadrant are added vectorially to yield the sum signal.

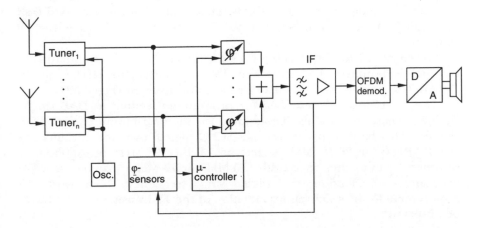

Fig. 4.31 Narrowband DAB receiver with antenna diversity (n = 2 to 4) (corresponding to ADA system for VHF FM from Blaupunkt)

Together with the criteria for digital audio broadcasting, such as efficient channel coding (1/2 rate), the guard interval and time and frequency interleaving, narrowband DAB has a good chance of meeting the quality criteria and of being put into practice.

4.6 Programme distribution of analogue and digital signals to terrestrial TV transmitters

In Germany conventional programme distribution to the analogue TV transmitter networks is the responsibility of DBP-Telekom and is fed through wired line networks or via microwave links; in the case of the latter the sound is transmitted digitally according to the MSTD method. Fringe areas are covered with the aid of TV transposers. The conventional TV transposers will turn into low-power transmitters processing signals from the satellites.

Similar to the programme distribution to VHF/FM and DAB transmitters, programme distribution to the analogue TV transmitter networks and to DVB transmitters (digital video broadcast, Section 4.7) in future will be mainly digital via satellite. For the necessary data transmission capacity, the quality level of the transmitted digital TV has to be considered: HDTV, EDTV, SDTV and LDTV (Section 3.8.1). On the other hand, the satellite transmission capacity for standard DSR is 20 Mbit/s, for advanced DSR with a 16-level modulation mode (8PSK/2AM or 16QAM) it is 40 Mbit/s, and by using the full transponder bandwidth it may reach up to 80 Mbit/s.

For programme distribution to analogue TV transmitters for PAL, SECAM or NTSC signals, SDTV quality (4.5 Mbit/s) should be assumed. The programmes should be distributed with a higher quality level, between SDTV and EDTV, i.e. approximately 7 Mbit/s. This means that with standard DSR two digital TV signals can be taken to TV transmitters of a particular TV standard.

For digital TV transmitters in EDTV quality (11 Mbit/s) a standard DSR channel is required. For terrestrial TV transmitters in HDTV quality (30 Mbit/s) double the DSR data rate has to be assumed (Fig. 4.32).

The SCPC method described for programme feeding to DAB single-frequency networks can also be adopted for the distribution of digital TV signals. The basic capacity for an SCPC signal may for instance be 10.24 Mbit/s or 20.48 Mbit/s. Instead of the 27 MHz transponder, a transponder of greater bandwidth (36 MHz or 70 MHz) may be used. For programme distribution to a digital single-frequency TV network the requirement for bit-synchronous radiation of the TV transmitters also has to be observed.

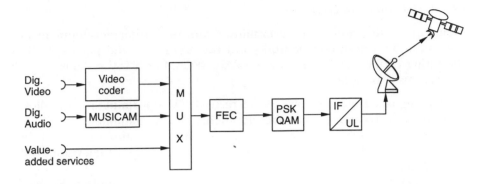

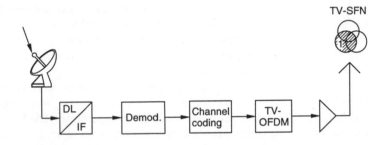

Fig. 4.32 Principle of digital programme distribution to digital and/or analogue TV transmitter networks (in parallel: direct satellite reception or reception via CATV possible)

		Mbit/s
On DSR basis:	4PSK	20
	8PSK	30
16QAM, 8PSK/2AM, 16PSK		40
2 DSR packets give double the capacity:		
(Line or microwave link)	32QAM	50
	64QAM	60

4.7 Digital video broadcasting (DVB)

The origination and processing of digital TV signals via satellite and CATV to the viewer is described as a model approach in Section 4.2. It is shown that the standard DSR method can be used for broadcasting up to EDTV quality, using existing transmission equipment including commercial DSR receivers with a special interface for 2×10.24 Mbit/s output. For HDTV quality, double the DSR channel capacity has to be provided as described in Section 4.3. With a given channel data rate, e.g. standard DSR of 20.48 Mbit/s, several programmes of SDTV or LDTV quality can be transmitted, the number of programmes being obtained through calculation.

4.7.1 Motivation for DVB

Some interesting aspects to be examined are the further development of terrestrial TV transmitter networks and the arguments and persuasion in favouring DVB instead of the present PAL, SECAM or NTSC systems.

Various reasons are:

- The signals from the present terrestrial TV transmitters only allow stationary reception. The development of the hand-held mini TV receivers (watchman) or the satellite and Yagi antennas used in motorhomes show that the interest in portable and mobile televisions will increase.
- Existing TV systems have reached the end of their technology S-curve, limiting the possibilities of further development, i.e. for value-added services.
- The applications of PALplus and D2MAC for terrestrial transmitters bring only a slight improvement and have compatibility problems.
- Subscriber numbers for terrestrial television in Germany are falling and it is expected that by 1995 more than 60% of homes would either be connected to cable or a satellite TV system.
- The frequencies of the existing channels are not economically distributed. For instance, about 20 TV channels are required for nationwide coverage of Germany's second TV programme [238].
- The terrestrial transmitter network is expensive in terms of power requirements. For a nationwide TV programme in Germany, for instance, 100 transmitters consuming approximately 3 MW in total are required.

4.7.2 Modulation methods

The basic parameters of DVB can be indirectly derived from the DAB specifications for modes 1, 2 and 3. A common feature of all three modes is the gross data rate of 2.4 Mbit/s (Table 2.9) per OFDM signal. DVB is based on a TV channel bandwidth of 7 MHz for VHF and 8 MHz for UHF. Thus a gross data rate of 9.6 Mbit/s is obtained for one TV channel in the VHF band with a configuration of four packets (Table 4.7). Assuming SDTV quality is with 4.5 Mbit/s and using channel coding at 1/2 rate, a single-frequency network for mobile reception could operate in band III. At the LDTV quality level three programmes could be transmitted for mobile reception. The mere possibility of receiving teletext in mobile mode is already an interesting application.

4.7.2.1 PSK/QAM

To obtain EDTV quality (approximately 11 Mbit/s) two TV channels are used, but it is preferable to choose a solution with a higher modulation depth. For DAB, QPSK modulation per carrier is specified. The 8PSK/2AM

DVB		Mode I	Mode II	Mode III
Frequency range		< 375 MHz	< 1.5 GHz	< 3 GHz
Application		SFN	local coverage	satellite
T_s total symbol period ($= t_s + \Delta$)		1.25 ms	312.5 μs	156.25 μs
t_s useful symbol period		1 ms	250 μs	125 μs
Δ guard interval		250 μs	62.5 μs	31.25 μs
Frequency bandwidth		Data carrier	Gaps	
B_{VHF}		4 · 1.536 MHz	+ 5 · 0.17 MHz =	7 MHz
B_{UHF}		5 · 1.536 MHz	+ 6 · 0.05 MHz =	8 MHz
B_{USA}		3 · 1.536 MHz	+ 4 · 0.35 MHz =	6 MHz
Number of carriers				
N_{VHF}		6144	1536	768
N_{UHF}		7680	1920	960
N_{USA}		4608	1152	576
Gross bit rate Mbit/s				
QPSK	7 MHz	9.6	9.6	−
	8 MHz	−	12.0	12.0
8PSK/2AM	7 MHz	19.2	19.2	−
or 16 QAM	8 MHz	−	24.0	24.0
	6 MHz	14.4	14.4	14.4
32QAM	7 MHz	24.0	24.0	−
	8 MHz	−	30.0	30.0
	6 MHz	18.0	18.0	18.0
64 QAM	7 MHz	28.8	28.8	−
	8 MHz	−	36.0	36.0
	6 MHz	21.6	21.6	21.6

Table 4.7 DVB parameters

and 16QAM methods described in Section 4.3, based on a single-carrier DSR, can also be employed in a multicarrier system. For 8PSK/2AM modulation, two DAB OFDM modulators operate with vector diagram co-ordinates rotated by 45° and simple addition of the output signals. This gives 8PSK per OFDM carrier with two amplitude levels A1 and A2 corresponding to the amplitude ratio of the two diagonals of an equilateral parallelogram with the angles of 45° and 135°. This is sent as an 8PSK/2AM signal and is derived from a circuit configuration with double IFFT, D/A conversion and I/Q modulation (Fig. 4.33).

An alternative method is 16QAM, which may also be implemented by two

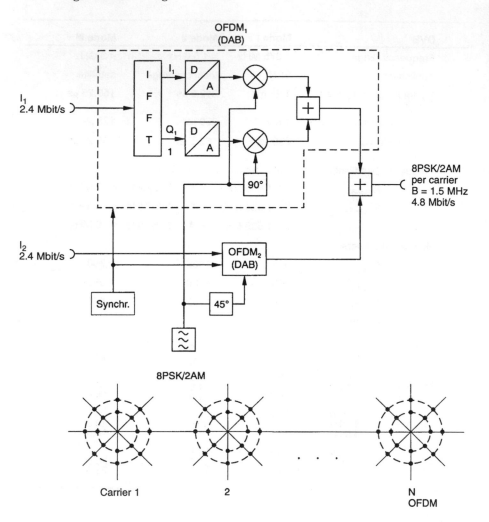

Fig. 4.33 8PSK/2AM multicarrier method

OFDM modulators, the output signal of one modulator being attenuated by 6 dB prior to the adder stage (Fig. 4.34).

A comparison of the two modulation methods is given in Section 4.3.3 and holds true also for their use in a multicarrier system.

When implementing 8PSK/2AM or 16QAM with two given OFDM modulators, the orthogonal arrangement of the carriers and the guard interval are maintained. The transmission bit rate is doubled and the decision circles are reduced accordingly.

With 16QAM, the decision ranges of the points within the modulation diagram are constant, whereas with 8PSK/2AM the decision circles for the

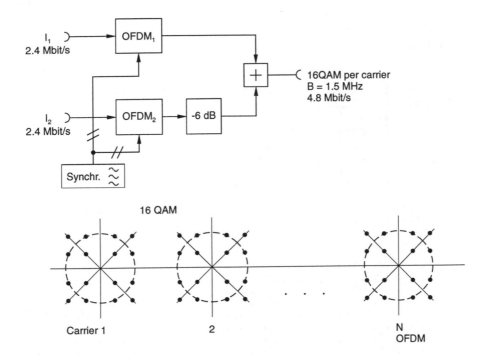

Fig. 4.34 16QAM multicarrier method

points on the outer circle are significantly larger than for the points on the inner circle. This difference can be compensated by an unequal error protection for the modulation points (Section 4.3.2). The method can, however, be used and even has the advantage of a graceful degradation.

For the terrestrial transmission of TV signals in HDTV quality, 32QAM or 64QAM should be used. 64QAM can be implemented with the aid of three standard OFDM modulators whose output signals can be added with a level ratio of $1:^1/_2:^1/_4$ (Fig. 4.35). Selecting attenuators of greater than 6 dB or 12 dB yields nonuniform 64QAM.

A data rate of 3×2.4 Mbit/s giving 7.2 Mbit/s per OFDM packet is obtained. In the UHF band, five OFDM packets can be arranged in the 8 MHz channel, yielding a total gross data rate of 36 Mbit/s. This data rate can be used for DVB in mode 2 and mode 3 according to Table 4.7. For stationary reception, HDTV quality with 25 to 30 Mbit/s baseband coding can be achieved while 32QAM is derived from 64QAM by selecting the modulation points (Fig. 4.36).

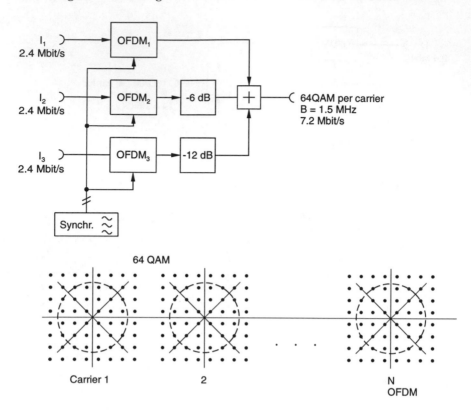

Fig. 4.35 64QAM multicarrier method

4.7.2.2 Trellis-coded modulation

TCM (trellis-coded modulation) is an approach to a modulation gain [237]. TCM is a combined coding and modulation technique used for digital transmission in band-limited channels. While in the transmitter a special finite-state coder is used if, for example, 8PSK is selected instead of QPSK, allowing only conditional phase shifts for PSK modulation, decoding in the receiver is performed by a soft-decision maximum-likelihood sequence decoder. With simple four-state TCM the susceptibility to additive noise can be improved by 3 dB against uncoded modulation. With a more complex TCM an improvement of 6 dB can be achieved with the same bandwidth and the same effective data rate.

Another way of using trellis modulation and Viterbi decoding for DVB is by adapting the modulator and to connect the Viterbi decoder to an appropriate point of the PSK/QAM demodulator in the receiver. Trellis coding in conjunction with OFDM is especially suitable for digital terrestrial television while the additional Viterbi decoder required for the receiver is more justifiable for TV than for DAB.

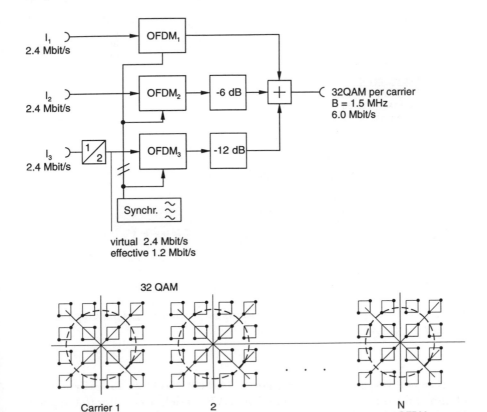

Fig. 4.36 32QAM multicarrier method, derived from 64QAM by using half the number of modulation points

4.7.3 Implementation of DVB models

Fig. 4.37 shows a basic approach to the hardware implementation of DVB models. On the basis of DAB multiplexes and modes 1, 2 and 3, three to five DAB packets of 1.536 MHz can be packed into one TV channel (6, 7 or 8 MHz bandwidth). The DAB packets cover in the time domain the entire symbol period T_S (= useful symbol period t_s + guard interval Δ). The type of modulation per carrier can be selected according to the DVB model to be implemented: QPSK (DAB), 8PSK/2AM or 16QAM, 32QAM, 64QAM. This implementation can be based on the formula:

$$\text{DVB} = m \times n \times \text{DAB packet (2.4 Mbit/s)}$$

where n is the number of packets in the TV channel (e.g. 3 to 5) and m the modulation depth based on the QPSK standard.

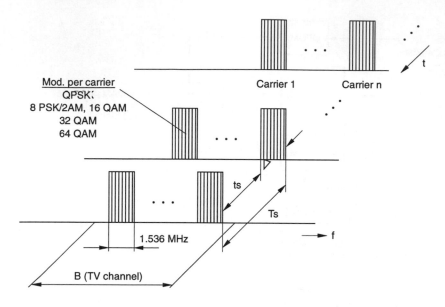

Fig. 4.37 DVB model with variable parameters on the basis of DAB multiplexes and modes 1, 2 and 3 (see also Table 4.7)

Using OFDM modules per packet (1.5 MHz), 3 to 15 modules in total are obtained for the implementation of the DVB models according to the given TV channel bandwidth and the chosen modulation depth.

As there are 1536 carriers in DAB mode 1, the basic module is based on a 2K IFFT, which is performed within the useful symbol period $t_s = 1$ ms (2K = 2048).

Note: With the aid of the Fourier transform a time domain signal is transformed into the frequency domain: since this is an integral transformation, it cannot be performed by a computer (Fig. 4.38). Computer calculation only becomes possible upon introduction of time-discrete samples.

The discrete Fourier transform uses a sum formula to transform a time interval determined by N samples into the frequency domain to give a discrete spectrum. The fast Fourier transform is an optimised fast Fourier algorithm and a special form of the discrete Fourier transform; the number of sampling points must be in a power of two series: $N = 2^x$ (e.g. 256, 512, 1024...). FFT processors are also provided with an IFFT algorithm.

The guard interval is obtained by extending the I/Q output registers of the IFFT processor by $1/4$ in time and by reading·them (Fig. 4.39). In the DAB modes 2 and 3 the same data volume is processed with a smaller number of carriers and reduced symbol period ($N \times 1/t_s$ = const.).

FT

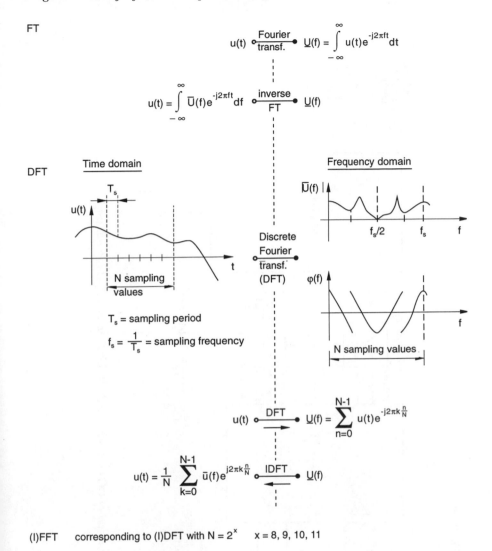

$$u(t) \, \overset{\text{Fourier}}{\underset{\text{transf.}}{\circ\!\!-\!\!\bullet}} \, \underline{U}(f) = \int\limits_{-\infty}^{\infty} u(t) e^{-j2\pi f t} dt$$

$$u(t) = \int\limits_{-\infty}^{\infty} \overline{U}(f) e^{-j2\pi f t} df \, \overset{\text{inverse}}{\underset{\text{FT}}{\circ\!\!-\!\!\bullet}} \, \underline{U}(f)$$

DFT

Time domain

Frequency domain

T_s = sampling period

$f_s = \dfrac{1}{T_s}$ = sampling frequency

Discrete Fourier transf. (DFT)

$$u(t) \, \overset{\text{DFT}}{\circ\!\!-\!\!\!\longrightarrow\!\!\bullet} \, \underline{U}(f) = \sum_{n=0}^{N-1} u(t) e^{-j2\pi k \frac{n}{N}}$$

$$u(t) = \frac{1}{N} \sum_{k=0}^{N-1} \bar{u}(f) e^{j2\pi k \frac{n}{N}} \, \overset{\text{IDFT}}{\circ\!\!\longleftarrow\!\!-\!\!\bullet} \, \underline{U}(f)$$

(I)FFT   corresponding to (I)DFT with $N = 2^x$ $x = 8, 9, 10, 11$

Fig. 4.38 Fast Fourier transform [244]

Therefore, it is assumed that the same IFFT processor can be used for all three modes.

QPSK is performed in the basic module, determining that all the 1536 occupied input registers for the real and imaginary components have the standard amplitude 1 and a phase of 0° or 180° depending on the bit position. The remaining input registers are set to 0.

Examples of suitable FFT chips range from the signal processor DSP

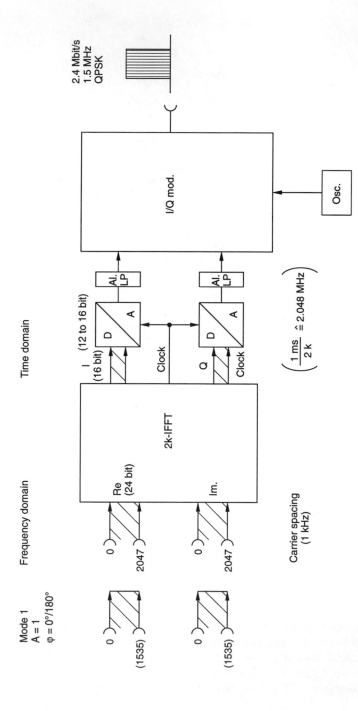

Fig. 4.39 Principle of OFDM module with 2K IFFT (mode 1) (Al-LP: aliasing lowpass filter)

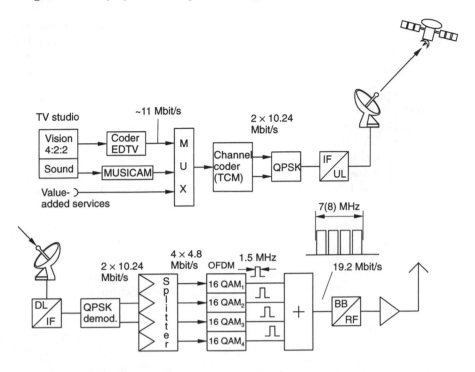

Fig. 4.40 Block diagram of terrestrial digital EDTV transmitter system with 16QAM (alternatively 8PSK/2AM) and with satellite distribution (DSR principle)

56001 from Motorola, the PDSP 16510 from Plessey, and the HDSP 66110 and HDSP 66210 from MicroARRAY (formerly Honeywell). A suitable digital/analogue convertor would be the HDAC 52160 from SPT featuring 16-bit resolution, but for this application 12 bit would suffice.

As the 2K IFFT only uses 1526 of the 2048 carriers an aliasing lowpass filter with noncritical skirt selectivity can be connected to the output of the D/A convertor. The aliasing filters for I and Q must, however, have the same characteristic. For the use of SAW filters, equal phase characteristics are also of key importance. A possible solution is adaptive equalisation with the aid of training sequences or the entry of the respective phase offset on the FFT chip.

The frequency position of the OFDM packet at the output of the basic module can be determined with the aid of the oscillator frequency of the I/Q modulator for simple addition and concatenation of several OFDM packets.

Fig. 4.40 shows an EDTV transmission system with 19.2 Mbit/s channel data rate which is based on the fundamental parameters of the tried and tested DAB (in this example 16QAM and 4 packets) and DSR methods. Since

16QAM is used instead of QPSK, it also has to be checked whether the mobile receiving capability has to be restricted. In the example described, channel coding for the terrestrial transmission is already performed at the earth station of the satellite link. The various arguments favouring this method are:

- The expected channel coding for terrestrial transmission largely eliminates the need for channel coding on the satellite link.
- Channel coding at the earth station renders this function superfluous in the many DVB transmitter stations.
- The requirement for bit-synchronous radiation is satisfied to a high degree *a priori*, since at the DVB transmitter only the 16QAM-OFDM modulator with a fixed algorithm for the bit allocation per carrier is required.

The upper limit of the data transmission capacity is reached with the system shown in Fig. 4.41. Five packets are arranged in the 8 MHz channel and 64QAM is employed, yielding a channel data rate of 36 Mbit/s ($3 \times 5 \times 2.4$ Mbit/s). For programme distribution, the DSR data capacity has to be doubled by using 8PSK/2AM, 16QAM or 16PSK modulation.

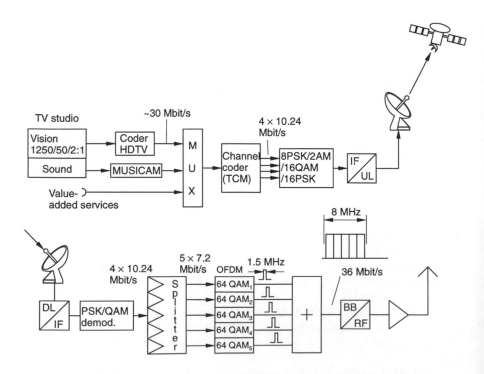

Fig. 4.41 Block diagram of HDTV system with 64QAM satellite distribution with 4-level modulation (4 bits per step)

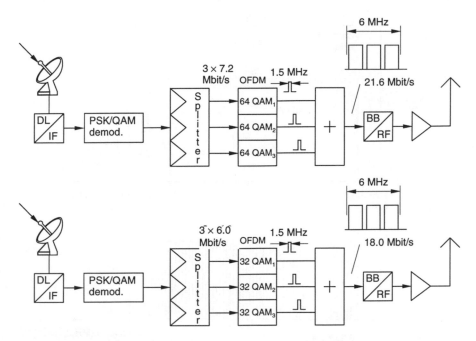

Fig. 4.42 Block diagram of terrestrial digital ADTV transmitter system for 6 MHz channel bandwidth (US) with 64QAM (top) and with 32QAM

Figs. 4.42 and 4.43 shows a solution with three OFDM modules for a terrestrial digital ADTV (advanced definition TV) transmitter system using a 6 MHz channel bandwidth (USA) with 64QAM and with 32QAM. The resulting channel data rate is 21.6 Mbit/s (64QAM) or 18.0 Mbit/s (32QAM). This approach is an alternative to the terrestrial digital HDTV system discussed and tested in the US (Section 3.8.2).

In view of a noncritical taboo channel compatibility of digital and analogue TV systems, it is an advantage, especially for the US system models, if no signal is transmitted at those points in the digital transmission channel at which there are distinct energy peaks in the analogue transmission channel (vision and sound carrier). These gaps could be produced by appropriate omission of the carriers in the first OFDM packet, which would reduce the data rate. It would also be advisable to use two modules for the first OFDM packet. Fig. 4.44 shows the arrangement of the OFDM modules in the 8 MHz channel with a total of four packets.

The previous models of implementation are based on an addition of analogue OFDM signals. This addition can successfully be digitally performed after the IFFT. Fig. 4.45 gives an example with 32QAM. For adding up the digital I/Q signals, adder chips (e.g. 16-bit TRW) are available, reducing the circuit to a pair of D/A convertors and an I/Q modulator.

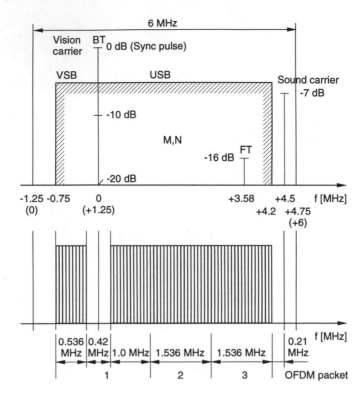

Fig. 4.43 TV channel occupied by OFDM packets for optimum adjacent-channel compatibility, shown for 6 MHz channel (US)

Instead of the previously assumed QPSK, the IFFT chip can also handle any modulation mode of a higher level (e.g. 8PSK/2AM, 16QAM, 32QAM or 64QAM). Upon each modulation step the complex modulation points have to be applied to the processor according to real and imaginary components. The modulation points may be stored in a ROM to be addressed by the input datastream.

The number of FFT chips required depends on the performance capacity of the FFT/DSP. In 1993, peak performance was offered by the Processor Type LH 9124 from Sharp (Table 4.8) [241]. It performs a complex 1K transformation in less than 100 µs and a complex 4K transformation in less than 400 µs. This allows an optimised hardware configuration as shown in Fig. 4.46. The preprocessor with LCA (logic cell array) supplies the complex modulation points for each carrier and for each modulation step. The processor performs an 8K-IFFT in two steps of 4K each with the results of the first transformation being buffered in an output register. At the end of the modulation step the equidistant samples of the time domain are simultaneously available. In this way, a DVB system with e.g. 24 Mbit/s gross data capacity can be configured with the optimised hardware.

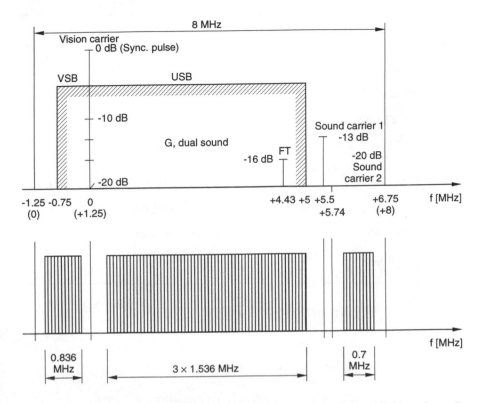

Fig. 4.44 8 MHz channel (UHF, Standard G) occupied by four OFDM packets of 1.536 MHz each (requires five OFDM modules)

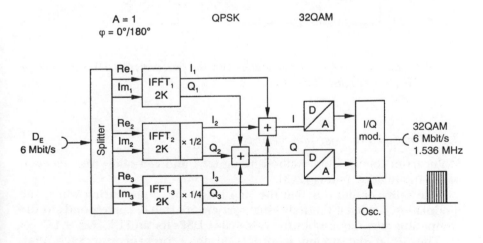

Fig. 4.45 Example of 32QAM implementation with addition of digital I/Q signals

Processor	Time	Accuracy	Factor
80386 (20 MHz)	200 ms	16-bit fixed point	1
TMS 320C25	15.8 ms	16-bit fixed point	12.6
TMS 320C30	2.5 ms	32-bit floating point	80
ADSP2100 (8 MHz)	7 ms	16-bit fixed point	28.6
M 56001	5 ms	24-bit fixed point	40
M 96001	2 ms	32-bit floating point	100
CRAY X-MP	1 ms	64-bit floating point	200
SHARP LH 9124 (40 MHz)	80 µs	24-bit fixed point	2500

Table 4.8 Benchmark for time required to perform complex 1K-FFT [241]

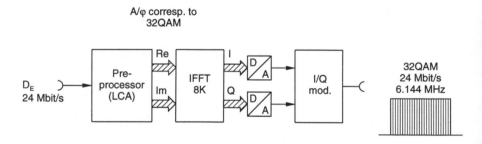

Fig. 4.46 Example of OFDM modulator with the parameters of DVB mode 1 for 24 Mbit/s per packet (Table 4.7)

4.7.4 Optimised transmission methods

Multicarrier transmission methods using IFFT and FFT allow free selection of the modulation points [243].

The only condition is that the step length, i.e. time during which the modulation points of the individual carriers are stable, corresponds to the computing time required by the cascadable DSPs for an FFT (Fig. 4.47).

The serial datastream is processed in a preprocessor such that, depending on the level of modulation (e.g. 16QAM), for the stored complex

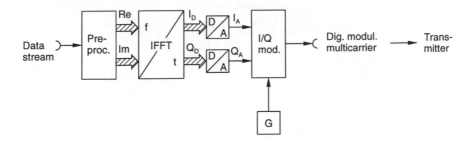

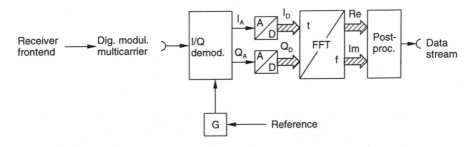

Fig. 4.47 Principle of digital modulation in multicarrier methods with free selection of modulation points (preprocessor)

modulation points are addressed according to real (Re) and imaginary (Im) components and taken to the IFFT processor.

The IFFT is performed along the step length and supplies while in the time domain the equidistant samples according to the Fourier transform law. The digital I and Q signals are D/A converted and then converted into the RF by an I/Q modulator.

Demodulation in the receiver (Fig. 4.47 bottom) starts with an I/Q demodulation using a referenced generator. The frequency and phase reference is contained in the modulation signal in time discrete form (reference symbol) or derived by averaging from the modulation signals. Due to the use of differential coding which may also be employed with modulation methods of higher order, there is no need to determine the absolute phase for a coherent demodulation.

After D/A conversion the signals IA and QA are taken to the FFT processor. The FFT algorithm gives the complex modulation points of the individual carriers within a multicarrier packet of equidistant frequencies. Interferences along the transmission channel cause displacement or dispersal of the modulation points within the decision circle. The latter is determined by the modulation mode selected. The postprocessor allocates

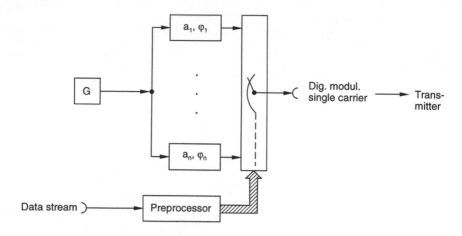

Fig. 4.48 Basic principle of digital modulation in single-carrier systems with free selection of modulation points (a, φ)

the modulation points received to the appropriate decision circles to determine the bit pattern. By combining the individual data of the carriers according to the step frequency the transmitted datastream is obtained.

Fig. 4.48 shows the basic principle of digital modulation in a single-carrier system with high data rate and free selection of the modulation points. The generator signal is taken via n paths — according to the modulation depth — in which a complex vector is formed according to magnitude (attenuation a) and phase (φ). A preprocessor controls the signal switch according to the available bit pattern and selects the vector or modulation point.

The free choice of the modulation points for a given modulation depth allows optimised modulation methods to be derived which are superior to the known methods. As shown in Fig. 4.49, 8PSK (top) can for instance be turned into 8POM (power optimised modulation). This reduces the average transmitter power p_r= 2.0 dB (37%) with decision circles remaining constant.

In the middle of Fig. 4.49 16QAM is compared with 16COM (circle optimised modulation). The area of the decision circle for the modulation points is increased by an f_v of approximately 35% with full driving being maintained. Turning 16QAM into 16POM yields a power reduction of p_r = 0.47 dB (10%).

The bottom of Fig. 4.49 shows the differences between 32-level ($f_v \approx 35\%$, p_r = 0.48 dB or 14%) and 64-level modulation ($f_v \approx 35\%$, p_r = 0.73 dB or 15%).

Instead of the above optimisation criteria (decision circle, power), combinations (COM/POM) or other parameters such as nonuniform decision circles (graceful degradation) may also be used.

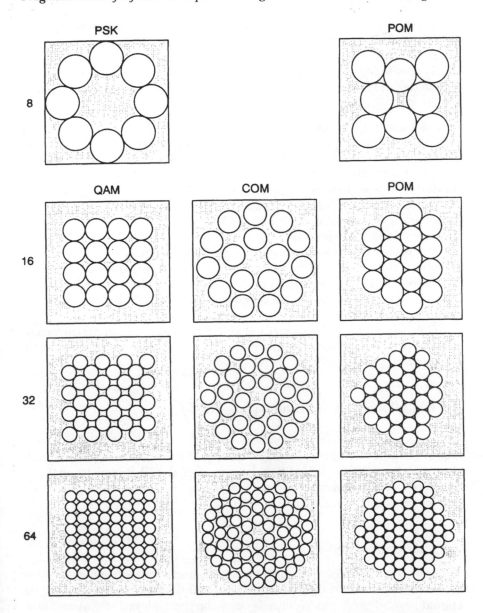

Fig. 4.49 New modulation modes such as COM and POM employed with 8, 16, 32 and 64 level modulation in comparison with PSK and QAM (with COM and POM the decision circles shown are identical with those of PSK/QAM; the actually larger decision circles of COM result from the utilisation of the increased centre point spacing)

The power reduction can also be considered as a modulation gain and may yield other advantages such as larger coverage areas, lower bit error rate or smaller size of receiving antenna.

By combining high grade channel coding for instance with trellis-coded modulation (TCM) and Viterbi coding, additional synergy effects can be obtained.

Note: A further approach to system optimisation is the OFDM transmission with constant envelope. This subject is treated in a research project of the Technical University of Darmstadt (Germany) [246]. The aim of this project is to find a suitable modulation to avoid the high voltage peaks that are inevitable with multicarrier signals and to achieve even a constant OFDM envelope. This would allow the linear amplifiers of the DAB transmitters to be driven up to about their nominal CW power with the benefit of lower investment and operating cost.

The basic concept is to use part of the carrier signals — e.g. every fifth carrier — not to be used for the transmission of information but for the compensation of envelope peaks by varying the amplitude and phase.

The compensation algorithm is first simulated with the aid of a computer and then implemented in real time for practical operation.

The disadvantage of this method is the reduction of data transmission capacity determined by the degree of higher level modulation, traded by the modulation gain by higher transmitter power.

4.8 Digital integrated broadcasting (DIB)

We have come full circle: the jigsaw pieces add up to a completed picture of digital generation, transmission and display of audio and video TV signals. Costly and quality impairing analogue to digital and digital to analogue interfaces are becoming obsolete. Only the interface to humans will (hopefully) remain analogue (Fig. 4.50).

4.8.1 Value-added services

The mosaics of digital sound and TV broadcasting also contain data or value-added services. The worlds of audio, video and data are rapidly converging. Digital sound and video signals may be considered as anonymous datastreams. To put it another way, there is a demand for transparent networks using different ways of transmission via terrestrial transmitters, satellite and cable with open interfaces to the various services. DAB and DVB will become DIB (digital integrated broadcasting). The terrestrial sound and TV broadcasting network will be turned into a BCN (broadcast communication network).

Fig.4.50 Newspeak!

Considerations for value-added services include:

- video on demand
- pay per view
- interactive TV (home shopping, education, games)
- traffic information service, traffic management/navigation
- defence against crimes (car theft, misuse of credit cards)
- updating of business data (stock exchange, lists, offers)
- mobile teletext and videotext
- radio paging services
- mobile fax
- special-type programmes (magazines, catalogues, telephone books, encyclopedias)
- pay radio.

Note: The example of teletext in the car clearly shows the DAB potential for value-added services.

While the information content of teletext is very comprehensive and widely accepted, the required data rate is relatively low. With 15 teletext lines in the frame (e.g. seven lines in the first field and eight lines in the second field) and 360 bits (45 bytes per sequence) in the active line period of 52 µs, the data rate for a continuous bitstream is 15×360 bits : 40 ms = 135 Kbit/s, little more than a MUSICAM audio mono channel!

In the future there will be hierarchical quality levels for audio and video already represented in audio by MUSICAM.

CD quality is obtained at 128 Kbit/s mono or with slight degradation at 96 Kbit/s mono. A data rate of 64 Kbit/s mono is suitable for feeding into the ISDN network, and for voice transmission a bit rate of 32 Kbit/s mono is appropriate.

The standardisation process for video was largely completed by the end of 1993. The top and bottom quality levels are HDTV and LDTV.

4.8.2 Display panel and mini LCD

An essential criterion for the acceptance of HDTV is the telepresence effect made possible by the large screen. The following technologies have been competing since 1993:

- *16:9 picture tube* with a screen diagonal measurement of 72 cm has been developed by Philips Components. The slimline design features a tube length of less than 62 cm and a weight of 35 kg (anode voltage 30 kV).
- The *LCD projector* from Sharp offers a 40–200-inch picture using TFT technology and an active matrix LCD. The video projector uses three 5.5-inch LCD panels for R, G, B, with each panel having 1.2 million pixels.
- The *flat panel display* from NHK (Japan Broadcasting Corporation Research Labs) has a 40-inch 16:9 format (picture size 0.87 m × 0.52 m), a depth of 8 cm and a weight of 8 kg. The hung-on-a-wall panel contains 1.1 million display cells which are driven by specially developed hybrid ICs. The PDP (plasma display panel) can display 256 colours. The new PDPs differ from previous versions by a pulse memory method (PMM), which ensures that the individual cells are active and light-emitting over long time periods [240].
- *Laser TV* was initiated in 1993 by Schneider AG in Germany. A laser beam is projected from the front or rear onto a projection area of up to several metres wide. The TV receiver featuring low power consumption and the dimensions of a video recorder complies with the 'ecovision' concept since it does not need an expensive, heavyweight vacuum picture tube that is such a problem for disposal. The first generation of models (with lasers similar to the ones in CD players) are scheduled to appear on the market in 1996.

In contrast to HDTV quality, a solution was also sought for those viewers who are mainly interested in the quantity of programmes offered rather what can be achieved technically regarding quality, and for those who use portable and mobile receivers, these desires already having led to the development of the hand-held watchman.

This mini-LCD TV set was offered in 1993 by eight manufacturers, although the existing TV transmitter network is hardly suitable for this use. There is an analogy to VHF sound broadcasting, which was originally designed for home reception but has gained more prominence through the car radio.

Portable LCD mini TV sets are offered in a variety of models and at favourable prices (some less than £100). There are special channel-compatible multistandard sets with on-screen displays, built-in telescopic antennas and automatic tuning. The picture resolution of the display is determined by the number of pixels and the LCD cells are arranged in a matrix with the position of each pixel defined by the column and line. A display of, for instance, 110 lines and 360 columns has 36,900 pixels which are addressed by 470 control lines [239]. The active matrix displays are implemented in TFT (thin-film transistor) technology.

4.8.3 Reserve potentials of the media

As in the USA, the trend in Europe will be to increase the number of programmes allocated to the same frequency. This situation will result because of the increasing number of private programme providers demanding more channels. An extra pressure is the internationalisation and diversification of the programmes caused by the advent of satellite broadcasting. The pressure for more channels may eventually lead to the position where up to four TV channels may be transmitted in standard PAL quality on a current TV channel.

Similar to mobile sound broadcasting, portable and mobile television will gain more popularity in the future. The main advantages of stationary TV reception are the large variety of programmes at the highest quality. The future for mobile reception will lie in its installation in coaches, taxis, luxury cars or motorhomes, on parking areas, camp sites and in traffic jams. Teletext capability on today's portable receivers is a growing requirement, and newscasts may also be of interest for car passengers or users of public transport.

The future development of the broadcasting media will be in accordance with their strength (Fig. 4.51). Digital technology will be optimised for every transmission medium, whether terrestrial, cable or satellite.

Reserve potentials of the transmission media may be used for digital transmission.

The coaxial cable has its reserve potential in the hyperband and in the lower UHF band. Coaxial and fibre optic cables will therefore have the advantage of being able to enhance the programme variety and offer several programmes at high quality simultaneously. By activating the return channel in the coaxial copper cable (5 to 20 MHz), systems for interactive communication can be implemented.

Programme variety and technical quality strengthen the argument in favour of satellite. Two recent circumstances have significantly contributed to the satellite as a programme distribution link to the consumer:

- The charges for using satellite receiving systems have been abolished in Austria and Germany.
- In mid-1993, ASTRA 1C, the third satellite of the ASTRA family, was launched into orbit over Zaire. Due to the co-positioning of this satellite, programme coverage equal to a cable network is possible from a single point in space. In conjunction with extremely sensitive LNCs (low noise convertors), 48 analogue TV programmes can be received by budget fixed 50 cm dish antennas, dispelling an old argument in favour of full-area cabling. ASTRA 1D, 1E and 1F, scheduled for operation in 1995, are to provide a capacity of more than 150 digital programmes.

With direct satellite broadcasting, an answer to complex copyright problems will have to be solved because of international broadcasting and reception of programmes.

The satellite links will also have to play an important role in digital sound and TV broadcasting as programme distribution services to digital terrestrial transmitter networks. It will mainly depend on the satellite owners and their rental fees how quickly the changeover from programme distribution via

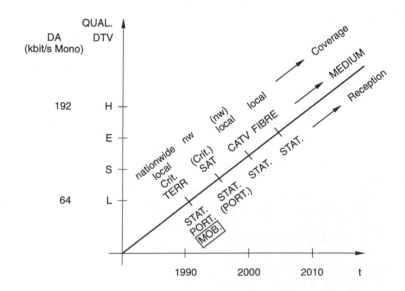

Fig. 4.51 Media in competition

terrestrial line networks and microwave channels to satellite distribution will be made.

Even the terrestrial transmitter network can offer a reserve potential. Digital television in HDTV quality for stationary reception and in LDTV/SDTV quality for mobile reception is planned for TV channels 61 to 69. These channels, which are currently reserved for NATO, are to be released. Transposing the TV channels of band I and/or band III into the UHF band (as in the UK, for instance) would release new frequency allocations. Another reserve potential is indirectly created by the change from conventional network planning to single-frequency networks.

The continual demand for the terrestrial transmitter networks is sustained due to the high area of coverage (99.5% in Switzerland and similar elsewhere in Europe) where networks can be installed and easily maintained even where coverage is partly critical, and where there is the need to supply homes with second or third TV sets using indoor antennas. The terrestrial medium is ideally suited for local and regional coverage, and when compared with satellite broadcasting, it offers higher security from unauthorised access and minimises the risk of a nationwide operational breakdown. The essential point in the longer term is that mobile reception of sound, picture and data is certainly in the domain of terrestrial broadcasting.

List of abbreviations

ABC	Annular beam control
ACATS	Advisory committee on advanced television service
ADA	Auto directional antenna
ADR	ASTRA digital radio
ADTV	Advanced definition TV
AES	Audio Engineering Society
AI	Auxiliary information
ALERT	Advice and problem location for European road traffic
ALU	Arithmetic logic unit
APL	Average picture level
ARI	Autofahrer-Rundfunk-Informationen (traffic information service for motorists)
ASCII	American standard code for information interchange
ASIC	Application specific integrated circuit
ASK	Amplitude shift keying
ASPEC	Adaptive spectral perceptual entropy coding
ATM	Asynchronous transfer mode
ATRAC	Adaptive transform acoustic coding
ATTC	Advanced television test centre
BB	Baseband
BCH	Block code according to Bose, Chaudhuri, Hocquenghem
BCN	Broadcast communication network
BER	Bit error rate
BK	Breitbandkommunikation (broadband communication)
BSS	Broadcast satellite services
BTS	Broadcast television systems
C/N	Carrier to noise
CAE	Computer aided engineering
CATV	Cable authority TV
CCD	Charge coupled device

CCIR	Comité Consultatif International de Radiodiffusion
CCETT	Centre Commun d'Etudes de Télédiffusion et Télécommunications
CCVS	Composite colour video signal
CD	Compact disc
CDR	Common data rate
CENELEC	Comité Européen de Normalisation Electrotechnique
CIRC	Cross interleave Reed Solomon code
CMOS	Complementary metal-oxide semiconductor
COFDM	Coded OFDM
COM	Circle optimised modulation
CPU	Central processor unit
CRC	Cyclic redundancy check
CW	Continuous wave
DAB	Digital audio broadcasting
DAT	Digital audio tape
DATV	Digitally assisted TV
DAVOS	Digital audio-video optical system
DB	Data broadcasting
DBS	Direct broadcasting satellite
DCC	Digital compact cassette
DCT	Discrete cosine transform
DCT	Digital component technology
DFS	Deutscher Fernmeldesatellit (German communications satellite)
DFT	Discrete Fourier transform
DIB	Digital integrated broadcasting
DMX	Digital music express
DPCM	Differential pulse code modulation
DQPSK	Differential quadrature PSK
DRAM	Dynamic random access memory
DSB-AM	Dual sideband amplitude modulation
DS1	Digital sound 1 Mbit/s
DSP	Digital signal processing
DSQ	Digital studio quality
DSR	Digital satellite radio
DTV	Digital TV
DVB	Digital video broadcasting
DVE	Digital video effects
E(I)RP	Equivalent (isotropic) radiated power
EBU	European Broadcasting Union (corresponds to UER)
ECC	Error correction code
ECS	European communication satellite

EDTV	Enhanced definition TV
EEPROM	Electrical erasable programmable ROM
ESA	European Space Agency
ESC	Energy saving collector
ETS	European telecom standard
ETSI	European Telecom Standardization Institute
EUREKA	European Research Commission Agency
EUTELSAT	European Telecommunication Satellite Organization
EWS	Emergency warning system
FCC	Federal Communications Commission
FEC	Forward error correction
FFT	Fast Fourier transform
FICS	Fully isolated coupling system (Thomson)
FIFO	First-in/First-out
FSS	Fixed satellite services
FTTC	Fibre to the curb
FTTH	Fibre to the home
FuBk	Fernsehausschuß der Funkbetriebskommission (Television Committee of the German Authority for television transmission)
GOPS	Giga operations per second
GPS	Global positioning system
GRP	Glassfibre reinforced plastics
GSM	Groupe Spécial Mobiles
HDB3	High density bipolar (maximum three consecutive zeros)
HDI	High definition interlaced
HDPCM	Hybrid DPCM
HDTV	High definition TV
IBC	International Broadcasting Convention
IDTV	Improved definition TV
IFFT	Inverse FFT
INTELSAT	International Telecommunication Satellite Organization
IOT	Inductive output tube
IRT	Institut für Rundfunktechnik (Institute for Broadcasting Technology, Munich)
ISDN	Integrated services digital network
ISO WG	International Standardization Organization Working Group
ITS	Insertion test signals
JESSI	Joint electron silicon semiconductor integration
JPEG	Joint photographic experts group

JTC	Joint Technical Committee (ETSI–EBU)
LCA	Logic cell array
LCD	Liquid crystal display
LDTV	Limited definition TV
LNC	Low noise convertor
LSB	Least significant bit
LSI	Large scale integration
MAC	Multiplex analogue component
MD	Mini disc
MDCT	Modified discrete cosine transformation
MJD	Modified Julian Day
MOD	Magneto-optical disc
MPEG	Motion pictures expert group
MSB	Most significant bit
MSC	Multiadaptive spectral audio coding
MSDC	Multi stage depressed collector
MSPS	Mega symbols per second
MUSE	Multiple subnyquist-sampling and encoding
MUSICAM	Masking pattern-adapted universal subband integrated coding and multiplexing
NB-QPSK	Narrowband quadrature PSK
NICAM	Near instantaneous companded audio multiplex
NMR	Noise-to-mask ratio
NRZ	Non return to zero
NTSC	National Television System Committee
OFDM	Orthogonal frequency division and multiplexing
PAD	Programme associated data
PAL	Phase alternation line
PASC	Precision adaptive sub-band coding
PCM	Pulse code modulation
PCN	Personal communication network
PDC	Programme delivery control
PDP	Plasma display panel
Pixel	Picture element (colour pixel, e.g. RGB)
POM	Power optimised modulation
PON	Passive optical network
PRBS	Pseudo-random bit sequence
PSK	Phase shift keying
QAM	Quadrature amplitude modulation

QMF	Quadrature mirror filter
RBDS	Radio broadcast data service (USA)
RDS	Radio data system
RISC	Reduced instruction set computing
RMS	Root mean square
RP	Radio paging
RS-,-	Reed Solomon . . . code
SAW	Surface acoustic waveform
SCA	Subsidiary channel authorisation
SCPC	Single channel per carrier
SDTV	Standard definition TV
SECAM	Séquentiel Couleurs à Memoire
SES	Société Européenne des Satellites
SFN	Single frequency network
SIS	Sound-in-sync
SMATV	Satellite master antenna TV
SMD	Surface mounted device
SS	Solid state
TCM	Trellis coded modulation
TFT	Thin film transistor
TMC	Traffic message channel
TWTA	Travelling wave tube amplifier
UEP	Unequal error protection
UER	Union Européenne de Radiodiffusion (corresponds to EBU)
UTC	Universal time co-ordinated
VCO	Voltage controlled oscillator
VCR	Video cassette recorder
VGA	Video graphics adaptor
VPS	Video programme system
VPT	Video recorder programming by teletext
VSB	Vestigial sideband (modulation)
VSWR	Voltage standing wave ratio
VTR	Video tape recording
WARC	World Administrative Radio Conference
4-DPSK	4-Differential PSK (corresponds to DQPSK)

References and sources

1 HUSCHKE, R., and MEHRGARDT, S.: 'Digitaltechnik in der Unterhaltungselektronik', *Digitale Signalverarbeitung im Fernsehgerät*, Moderne Industrie, 1990, **48**

2 BAUER, H.: 10 kW 'Tubed transmitter NR410R for VHF FM sound broadcasting', *News from Rohde & Schwarz*, 1989, **124,** pp. 25– 27

3 STEEN, R., and QUIRMBACH, B.: '10 kW tubed FM standby transmitter', *News from Rohde & Schwarz*, 1991, **135,** p. 45

4 LINCKELMANN, G.: 'Schaltung zum Aufteilen oder Zusammenführen von HF-Leistung'. German Patent Specification 27 33 888 (Rohde & Schwarz patent)

5 SEEBERGER, H.: 'Transistorized 10 kW VHF FM transmitter NR410T1', *News from Rohde & Schwarz*, 1992, **136,** pp. 15–17

6 BAUER, H.: 'Compact VHF FM transmitter with output power of 1.5, 3 or 5 kW', *News from Rohde & Schwarz*, 1984, **109,** pp. 16–20

7 DIETL, A., and WENDL, A.: 'VHF FM transmitter SU 115: exciter for VHF sound broadcasting', *News from Rohde & Schwarz*, 1984, **107,** pp. 20–22

8 SEEBERGER, H.: 'Leitungen, Transformatoren, Leistungsaddierer', *Training documentation for VHF FM transmitters*, December 1989

9 DALISDA, U.: 'Schaltung zum Aufteilen oder Zusammenführung von Hochfrequenzleistung'. Rohde & Schwarz Patent Specification 1251-P

10 BOHLEN, H.: 'The development of TV klystron efficiency: collector depression, advanced tuning, beam modulation', *IEEE Colloquium Digest*, 1984, **22**

11 ASCHERL, P., and Kislinger K.: '20 kW UHF TV Transmitter NT424', *News from Rohde & Schwarz*, 1984, **106,** pp. 31–33

12 ASCHERL, P.: '20-kW UHF TV transmitter NT425', *News from Rohde & Schwarz*, 1987, **118,** pp. 17–18

13 NN: 'UHF TV transmitter systems', *Rohde & Schwarz Info.*, N4-015 E-1

14 NN: 'Description of TV transmitter NT424 with ABC unit GX424, 20 kW/Band IV/V'. Rohde & Schwarz

15 HERWERTH, R.: 'Verfahren und Anordnung zum Aussenden von PAL-Plus-Signalen, Rohde & Schwarz', employee invention. P4304746.7 (1335-P), 2.3.1993

16 NN: 'TV transmitter system 2100 10-kW TV transmitter NT413S for Band III (170 to 230 MHz)', Special print PD 756.3124.11. Rohde & Schwarz 691(MHI dr)

17 NN: '10 kW UHF solid state transmitter'. Thomson LGT, France, data sheet

18 NN: '5 kW–10 kW UHF solid state transmitters'. Thomson LGT, France, data sheet issued November 1989

19 NIES, J.: 'Entwicklung des TV-Marktes in der BRD seit der Einführung privater Programmanbieter'. Rohde & Schwarz study in co-operation with K. Kislinger and P. Dambacher; 2AVP-Ni-je 25th June 1992
20 FREEMAN, D.B.: 'The flexible use of new high efficiency power amplification systems'. Television Technology Corporation USA, NAB 1992 Broadcast Engineering Conference Proceedings, pp. 18–21
21 GOEFFREY, T. C., BOHLEN, H.P., and HEPPINSTALL, R.: 'Some exciting adventures in the IOT business'. EEV Limited, Chelmsford, England, NAB 1992 Broadcast Engineering Conference Proceedings, pp. 200–208
22 WOZNIAK, J.: 'Using tetrode power amplifiers in high power UHF TV transmitters'. Acrodyne Industries Inc., Blue Bell, Pennsylvania, NAB 1992 Broadcast Engineering Conference Proceedings, pp. 209–215
23 McCUNE, E.W.: 'MSDC klystron field performance varian microwave tube products'. Palo Alto, California, NAB 1992 Broadcast Engineering Conference Proceedings, pp. 183–186
24 MÄUSL, R.: 'Television technology; satellite TV broadcasting', *R & S refresher topics*, *VIII*, pp. 32–34
25 WEINLEIN, W.: 'Überblick über den Stand der Satellitenrundfunktechnik', *Fernseh- und Kinotechnik*,1984, **38**, (6) pp. 239–250
26 PAULI, P.: 'Grundlagen für den Satelliten-, Fernseh- und Hörfunkempfang'. Training course SAT1, November 1985, Rohde & Schwarz advanced training programme
27 NN: *Kabel & Satellit*, June 1992, pp. 30–32
28 BIRKILL, S.: 'Transponder watch', datafile, *Cable and Satellite*, May 1993, p. 76
29 NN: 'Broadband communication: Rohde & Schwarz operations and test equipment', *News from Rohde & Schwarz*, 1985, **108**, pp. 39–42
30 KAISER, W.: 'Auf dem Weg zu Teilnehmeranschlüssen in Glasfasertechnik', Telecommunications **16**, Münchner Kreis, Glasfaser bis ins haus', pp. 14–36 (W. Kaiser, Springer-Verlag)
31 ZEIDLER, G.: 'Stand und Entwicklungstendenzen bei Lichtwellenleitern und den dazugehörigen passiven Baulelementen', Telecommunications, **16**, Münchner Kreis, Glasfaser bis ins Haus, pp. 39–52 (W. Kaiser, Springer-Verlag)
32 HEIDEMANN, R.: 'Entwicklungslinien optischer Weitverkehrssysteme und Komponenten', Telecommunications 16, Münchner Kreis, Glasfaser bis ins Haus, pp. 55–68 (W. Kaiser, Springer-Verlag)
33 SIEGLE, G.: 'Transparente Übertragung breitbandiger Signale über Monomode-Glasfasern', Telecommunications 16, Münchner Kreis, Glasfaser bis ins Haus, pp. 85–93 (W. Kaiser, Springer-Verlag)
34 NN: 'Berkom-Telekommunikation im Glasfasernetz', *DETECON Technisches Zentrum*, Berlin, 1990, issue 3
35 ZANDER, H.: DCC and MD: 'Die neuen Medien der digitalen Audiotechnik', *Fernseh- und Kinotechnik*, 1992, **46**, (10), pp. 680–686
36 NN: 'Magneto-optischer Discrecorder (MOD)', Research & Development Laboratories, Thomson Consumer Electronics Communications & PR Villingen, Germany
37 POHLMANN, K.C.: 'Principles of digital audio, B2/10449', pp. 215–225, Howard W. Sams & Co., Inc., Indianapolis, Indiana USA
38 HARDER, C.: 'Compact disc, digital audio', *Funkschau*, 1981, **18**, pp. 69–71
39 NN: 'CD-I, Mehr als nur Telespiele für Erwachsene', *Professional Media*, **92**, pp. 42–43
40 NN: 'Blauer Laser-Strahl liest dreimal schneller', *News from Reuter in Süddeutsche Zeitung*, July, 1992

41 THOMSEN, D.:'Digital audio tape (DAT): sechzig Meter Band – zwei Stunden Musik', *Funkschau*, 1987, **7**, pp. 28–31, (based on a lecture held at sound mixer conference 1986 in Munich)

42 NN: 'Digitale compact cassette', Première at photokina, *Professional Media*, 1992, pp. 52– 56

43 NN: 'DCC – Unterstützung auf breiter Front', *Funkschau*, 1991, **20**, p. 14

44 NN: 'Mini Disc', Funkschau, 1992, **15**, pp. 43–49

45 NN: 'Die Daten eingedampft ', *Funkschau*, 1992, **15**

46 NN:'EBU: Specifications of the radio data system RDS for VHF/FM sound broadcasting', *Tech. 3244-E*, March 1984

47 MIELKE, J.:'Die Übertragung von Zusatzinformationen im UKW-Hörrundfunk', *RTM 28*, 1984, **2**, pp. 69–73

48 DAMBACHER, P.: 'Radio data – a new service in VHF sound broadcasting', *News from Rohde & Schwarz*, 1984, **107**

49 MAYR, J.: 'So kommt RDS zum Sender', *Funkschau*, 1987, **16**

50 MAYR, J.: 'FM radio data decoder DMDC for radio data and traffic radio', *News from Rohde & Schwarz*, 1989, 127

51 MIESLINGER, H., and ZUREK, G.: 'RDS monitoring with FM radio data decoder DMDC05', *News from Rohde & Schwarz*, 1992, **137**, pp. 18–22

52 NN: 'ARD-Pflichtenheft: Betriebs- und Überwachungsdecoder für das Radio-Daten-System (RDS), Zusatzinformationen im UKW-Hörrundfunk', *UER Doc. Tech. 3244*, 1988, **5/3.9**, (1), Institut für Rundfunktechnik GmbH

53 LITZA, W.: 'RDS wonderland', *Rohde & Schwarz leaflet PD 756.8955.21*, 1991 (PAN dr)

54 NN: 'Nachrichtenverbundsystem für den Verkehrsfunk'. ST-SNÜ/9.91; Südwestfunk

55 NN: 'Jedem sein Verkehrsfunk', *Süddeutsche Zeitung*, 1992, **149**, p. 45

56 NN: 'FM radio data coder DMC05/09', *Rohde & Schwarz Manual*, description of front-panel controls, Ident No. 812.1310.02

57 KRÜGER, G., and Dr. SPRINGER, R.: 'Highly precise localization on land, at sea and in the air with GPS and DGPS', *News from Rohde & Schwarz*, 1993, **140**, pp. 26–27

58 TREYTL, P.: 'Digitaler Hörfunk über Rundfunksatelliten'. (Deutsche Forschungs- und Versuchsanstalt für Luft und Raumfahrt, Cologne)

59 DAMBACHER, P., and KRAHMER, E.: 'Audio coder DCA for digital sound broadcasting', *News from Rohde & Schwarz*, 1986, **114**, pp. 13–16

60 NN: 'Digital satellite radio (DSR), specification for the transmission method', *Technische Richtlinie Nr. 3R1 der öffentlich-rechtlichen Rundfunkanstalten in der BRD*, 1989, **Issue 3**, IRT

61 DIETL, A.: 'SFSP – test generator for digital satellite broadcasting', *News from Rohde & Schwarz,*' 1988, **122**, pp. 11–14

62 SCHMIDT, P.: 'Audio decoder DDA for digital sound transmission', *News from Rohde & Schwarz*, 1988, **122**, pp. 15–18

63 WALTER, W.: 'Digitale Tonübertragung für den Programmaustausch und für den Satelliten-Rundfunk', *Taschenbuch der Telekom-Praxis*, 1991, pp. 128–165

64 NN: 'Zusatzinformationsübertragung auf der DS1-strecke ARD/ZDF: Technische Richtlinie 3R2'. (Institut für Rundfunktechnik, Munich, December 1984)

65 NN: 'Schnittstellen-Kennwerte für digitale Stereosignale der Bitrate 1024 kbit/s (DS1)', *FTZ-Richtlinie 154R4*, August 1985

66 NN: 'Specification of the digital audio interface', *EBU-Tech. 3250*

67 NN: 'Supersound aus dem Weltall', *Funkschau*, 1991, **23**, pp. 65–67

68 NN: 'Informationen über Radio und Television', *IRT*, 1992, **78,** 3084 Wabern, Switzerland
69 NN: 'Radio via satellit', *Funkschau*, 1991, **23,** pp. 68–70
70 KRIEBEL, H., and Kleine, G.: 'Digitaler Satellitenrundfunk CD-Qualität aus dem Orbit', *Funkschau*, 1992, **19,** pp. 68–71
71 DAMBACHER, P., and SÜVERKRÜBBE, R.: 'Euroradio – the listening counterpart to Eurovision', *News from Rohde & Schwarz*, 1990, **128**
72 SÜBERKRÜBBE, R.: 'The DS1-system for Euroradio transmission', extract from system documentation, *IRT*, April 1988
73 NN: 'Digital studio quality (DSQ) sound-programme transmission equipment'. AT&T Network Systems AG Zurich, Technical Information 5TR620
74 KRAHÉ, D.: 'Ein Verfahren zur Datenreduktion bei digitalen Audiosignalen unter Ausnutzung psychoakustischer Phänomene', *Rundfunkt*, **30,** 1986, pp. 117–123
75 ZWICKER, E.: 'Psychoakustik', (Springer 1982)
76 NN: 'MUSICAM: a universal subband coding system description', *CCITT, IRT*, Matsushita, Philips, Annex 3
77 STOLL, G., WIESE, D., and LINK, M.: 'MUSICAM: Ein Quellencodierverfahren zur Datenreduktion hochqualitativer Audiosignale für universelle Anwendung im Bereich der digitalen Tonübertragung und -speicherung'. Taschenbuch der telekompraxis 1991, *Fachverlag Schiele & Schön*, **28,** pp. 96–127, (Berlin Schiele & Schön GmbH)
78 RENNER, S.: 'Daten-Diät Datenreduktion bei digitalisierten Audiosignalen', *Elrad*, 1991, **Issue 4**
79 SCHLÜTER, K.: 'Datenreduktion wie Radio und Telefon transparent werden', *Funkschau*, 1992, **1**
80 STOLL, G., and BRANDENBURG, K.: 'Das ISO/MPEG-Audio Codec, ein generischer Standard für die Codierung von hochqualitativen digitalen Tonsignalen', *IRT*. University Erlangen-Nuremberg/FhG. Script for 9th ITG Sound Broadcasting Conference, 18.-1992 in Mannheim
81 BRANDENBURG, K., HERRE, J., JOHNSTON, J.D., MAHIEUX, J., and SCHROEDER, E. F.: 'ASPEC: adaptive spectral perceptual entropy coding of high quality music signals'. 90th AES-convention, Paris 1991, preprint 3011
82 STOLL, G., and THEILE, G.: 'MASCAM: minimale Datenrate durch Berücksichtigung der Gehöreigenschaften bei der Codierung hochwertiger Tonsignale', *FKTG*, **42,** (11), pp. 551–558
83 STOLL, G.: 'Source coding for DAB and the evaluation of its performance: a major application of the new ISO audio coding standard', *IRT*. 1st International Symposium on DAB – 1992 Proceedings, pp. 83–97
84 NN: 'MASC3000 multipurpose audio signal compander', *Preliminary Product Information ITT Semiconductors*, March 1992,
85 DAMBACHER, P.: 'Finding the path to DAB', *Radio World*, **27,** December 1991 and further issues
86 PLENGE, G.: 'DAB – ein neues Rundfunkhörsystem, Stand der Entwicklung und Wege zu seiner Einführung', *Rundfunktechnische Mitteilungen*, 1991, **2,** pp. 45–66
87 MÜLLER-RÖMER, F.: 'Digitale terrestrische Sendernetze für Hörfunk und Fernsehen', *Fernseh- und Kinotechnik*, 1991, **11,** pp. 575–583
88 LE FlOCH, B., HALBERT-LASALLE, R., and CASTELAIN, D.: 'Digital sound broadcasting to mobile receivers', *IEEE Trans. on Consumer Electronics*, 1989, **35,** (3), pp. 493–503
89 WOLF, J.: 'DAB competence of Rohde & Schwarz', Sales Circular March 1992 (wf)

90 SCHNEEBERGER, G.: 'DAB-digitaler terrestrischer Hörfunk für mobilen Empfang', *Taschenbuch der telekom praxis*, 1992, pp. 148–185

91 NN: 'EUREKA 147/DAB system description DT/05959-III/C', December 1992 (general information for EIA testing procedure)

92 NN: Additional and updated information on digital system A (= DAB, developed by Eureka 147 Consortium and supported by EBU) EBU Document WP 10B/.., WP 10-115/.., December 1992

93 WOLF, J., and WINTER, A.: 'DAB-Meßverfahren und Meßkonzepte'. Proceedings of 4th meeting of Ad-hoc Working Group of DAB Platform, 1992, Chairman: P. Dambacher

94 MÄUSL, R.: 'Modulationsverfahren in der Nachrichtentechnik – ein Repititorium'. Rohde & Schwarz advanced training programme

95 LE FlOCH, B.: 'Channel coding and modulation for DAB CCETT', 1st International Symposiom on DAB, *Proceedings*, 1992, pp. 99–108

96 SCHNEEBERGER, G.: 'DAB-Statusreport', *IRT*, May 1992

97 Heinemann, C.: 'Der Dopplereffekt in Zusammenhang mit DAB', *Rohde & Schwarz study*, July 1992 in co-operation with P. Dambacher

98 NN: Technical information on DAB filters by Messrs. Spinner

99 SCHWAIGER, K.H.: 'Programm- und Datenzuführung zu DAB-Gleichwellensendernetzen', *Technical Report B 129/92*, Institut für Rundfunktechnik, 10/1992

100 KUHL, K.H., PETKE, G., and SCHNEEBERGER, G.: 'Zur Gestaltung von Sendernetzen für einen digitalen terrestrischen Hörfunk', *IRT*, **8**, October 1990

101 NN: 'Antennensysteme und Weichen für DAB im Kanal 12', Technical Information from Messrs. Kathrein, Antennen Electronic, September 1992

102 NN: 'DAB field trials in France', Documents CCIR Study groups period 1990–1994, Document 10B/15-E only, November 1992, Subject Question 51/10 Recommendation 774 Report 1203

103 MÜLLER-RÖMER, F.: 'Wie wird Digital Audio Broadcasting (DAB) die Rundfunklandschaft verändern?', Lecture within Panel 3: Neue Übertragungs, Programm- und Marketingformen für das Radio 2000: Digital Audio Broadcasting (DAB), DAB Platform, 8 October 1992

104 KADEL, G., and LORENZ, R.W.: 'Mobile propagation measurements using a digital channel sounder with a bandwidth matched to the GSM system', International Conference on Antennas and Propagation, *IEEE Conference Publication 333*, 1991, **1**, pp. 496– 499

105 WOLF, J.: 'DAB-Meßverfahren und Meßkonzepte', Proceedings of 1st and 2nd meeting of Ad-hoc Working Group of DAB Platform. The working group comprises representatives from broadcasters, industry and institutes. Chairman of meetings in November 1991 and February.1992: P. Dambacher

106 NN: 'ACORN DAB' *Technical Information*, NAB Las Vegas, 1992

107 NN: SES-ASTRA-ADR, MPM/93-F142B, CD, MPM/93-F115B

108 DAMBACHER, P.: 'Fortentwicklung der Fernseh-Meß- und Überwachungstechnik durch Prüfzeilensignale', *Fernmeldepraxis*, 1976, **53**, (12), pp. 548–554. *Fernmeldepraxis*, 1976, **53**, (14), pp. 627–654

109 DAMBACHER, P., and HENSCHKE, W.: 'Fernsehmeßtechnik mit Prüfzeilensignalen', *Funkschau*, 1974, 8, pp. 784–790

110 NN: CCIR-Recommendation 473-2 and CCIR Report 314-4

111 HARM, H., DAMBACHER, P., and HENSCHKE, W.: 'Abtast- und Halte-Schaltung zum Ermitteln des in einem periodischen Signal enthaltenen Spannungswertes'. Patent Specification DE 2441 192 C2. Date of disclosure 3rd June 1982. Patent holder: Rohde & Schwarz

112 NN: 'Schlußbericht der Ad-hoc-Gruppe Datenzeile des 'ARD/ZDF/DBP-Fernsehleitungsausschusses', *FELA*, November 1980

113 NN: Prüfzeilenmeßtechnik. Die Übertragung von 'Datenzeilen' im Fernsehsignal, FTZ-Ref. N44-4. DBP-Richtlinie FTZ 155 R 170 Teil 3

114 NN: 'Specification of insertion data signal equipment for international transmissions', *EBU*, 3rd ed. June 1977, Techn. 3217-E.

115 DESCHLER, W.: 'TV data line decoder DEF', *News from Rohde & Schwarz*, 1984, **108**, pp. 15–17

116 DESCHLER, W.: 'DGF, a coder for the TV data line', *News from Rohde & Schwarz*, 1986, **113**, pp. 17–19

117 SOMMERHÄUSER, W.: 'Flexibel programmieren mit VPS', *Funkschau*, 1985, **25**, pp. 47–51

118 EITZ, KOCH, Oberlies: 'Video-Programm-System', ARD/ZDF/ZVEI-Richtlinie, Teil VII – VPS 104, Fernsehtext programmiert Fernseh-Heimgeräte. IRT/ZDF/WDR March 1986

119 DERNEDDE-JESSEN, H.: 'Service über Videotext', *Funkschau*, 1986, **11**, pp. 40–43

120 NN: 'Transmission of several sound or information channels in television', *CCIR-Report 795*, **10**, Kyoto 1978, pp. 100–107

121 NN: 'Übertragung mehrerer Ton- oder Informationskanäle beim Fernsehen, CCIR-Studienkommission', *Bericht zum Report 795*, Bundesrepublik Deutschland, St.K10, May 1980

122 AIGNER, M., and GOROL, R.: 'Eine neue Stereomatrizierung für den Fernsehton', *Rundfunktechnische Mitteilungen*, 1979, **23**, (1), pp. 10–13

123 BURKHARDT, R., and STEUDEL, G.: 'Integrierte digitale Stereoübertragung im Fernsehen', *Rundfunktechnische Mitteilungen*, 1980, **24**, (1), p. 26

124 DIETZ, U., DINSEL, S., and OBERNDORFER, R.: 'Vergleichende Ausbreitungsmessungen im Fernsehbereich mit dem Zwei-Tonträger-Verfahren und dem FM/FM-Multiplexverfahren', *Rundfunktechnische Mitteilungen*, 1973, **17**, (4), p. 175

125 VOGT, N.: 'Mehrkanalton im Fernsehen', *Taschenbuch der Fernmelde-Praxis*, 1981, p. 196

126 AIGNER, M.: 'Zweitonübertragung beim Fernsehen. Der Einfluß des Offsetbetriebes von Fernsehsendern auf den Tonstörabstand beim FM/FM-Multiplexverfahren und beim Zweiträgerverfahren', *Rundfunktechnische Mitteilungen*, 1978, **22**, (4), pp. 185

127 DAMBACHER, P.: 'Stereo and dual sound in television', *News from Rohde & Schwarz*, 1981, **94**, pp. 9–13

128 ADLER, E.: 'Zwei Tonkanäle hoher Qualität', *Funkschau*, 1974, **11**, pp. 396–398

129 DAMBACHER, P., and SINGERL, P.: 'Aufwertung: Zweitonverfahren für das Fernsehen', *Funkschau*, 1981, **17**, pp. 73–77

130 DWUZET H.: 'Kombinierte Bild-Ton-Übertragung nach dem SIS-Verfahren', *Funkschau*, 1974, **23**, pp. 891–893

131 SHORTER: 'The distribution of television sound by pulsecode modulation signals in the video waveform', *BBC, EBU Review*, 1969, **113**, pp. 13–16

132 HILPERT,T., and HUSCHKE, R.: 'Digitaler Multistandard-Audio-Chipsatz', Special print from *Elektronik*, 1990, **6**, 16.3., Kennz. 6200-210-1D, ITT *Intermetall*, pp. 1–7

133 GEIER, G., and HERMSDÖRFER, O.: 'Measurement of TV digital sound NICAM-728', *News from Rohde & Schwarz*, 1989, **127**, pp. 40–41

134 NN: 'Digital sound transmission in terrestrial television', *EBU Technical Recommendation, 4.87 SPB 424 Revised Version*, Ref. CT/III-A mcdb

135 NN: 'NICAM 728: specification for two additional digital sound channels with system I television', BBC/IBA/BREMA Joint Publication (revised edition), August 1988, ISBN 0 563 20716 7

136 NN: 'Specification for transmission of two-channel digital sound with terrestrial television systems B, G and I', *EBU SPB 424*, Technical Center, Brussels

137 WESTERKAMP, D., and RIEMANN, U.: 'PAlplus: Übertragung von 16:9-Bildern im terrestrischen PAL-Kanal'. Lecture at the 15th annual conference of FKTG in Berlin, June 1992, Proceedings Part 1, Fernseh- und Kinotechnische Gesellschaft FKTG e.V.

138 MÄUSL, R.: 'PALplus – an improved PAL system',. *Television technology, R&S refresher topics IX/11.*, pp. 35–38

139 KRAMER, D.: 'PALplus – ein gangbarer Weg in die Fernsehzukunft', *Bulletin SEV/VSE81*, 1990, **21**, 10th November, ITG-sponsored conference on HDTV at SRG (Schweizerischer Rundfunkgesellschaft)

140 HENTSCHEL, C.: 'Bandaufspaltung zur Rasterkonversion für eine PALplus-Übertragung', *Fernseh- und Kinotechnik*, 1992, **45**, (11) p. 742

141 EBNER, A., MATZEL, E., MORCOM, R., OCHS, R., RIEMANN, U., SILVERBERG, M., STOREY R., VREESWIJK, F., and WESTERKAMP, D.: 'PALplus: Übertragung von 16:9-Bildern im terrestrischen PAL-Kanal', *Fernseh-und Kinotechnik*, 1992, **46**, (11), pp. 733–739

142 NN: 'D2-MAC/Packet für DBS', *Fuba Communication*, Hans Kolbe & Co., Bad Salzdetfurth

143 MESSERSCHMID, U.: 'Fernsehnormen für Rundfunksatelliten Das D2-MAC/Packet-Verfahren in der MAC-Systemfamilie', *Fernseh- und Kinotechnik*, 1985, **39**, (10)

144 NN: 'Television standards for the broadcasting satellite service – specification of the C-MAC/packet system', *EBU, Doc SPB 284*, 4th rev. version, Brussels, February 1985

145 NN: 'Methods of conveying C-MAC/PACKET signals in small and large community antenna and cable network installations', *EBU*, Chapt. A: Specification of the C-MAC/Packet System. Chapt. B: Specification of the D2-MAC/ Packet System. Doc. SPB 352, rev. version, Brussels, February 1985

146 NN: 'Frequency modulation parameters of the D2-MAC/packet system for DBS', *EBU*, Doc. SPB 368, Brussels, April 1985

147 MÄUSL, R.: 'MAC system television technology', *R&S refresher topics VIII/9.*, pp. 27–32

148 NN: 'Specification of the HD-MAC/packet system', contribution of the EUREKA-95 Project ETSI-Doc.22, February 1991

149 MÄUSL, R.: 'High-definition television (HDTV) television technology', *R&S refresher topics VIII/8*, pp. 25–27

150 VREESWIJK, F., and Haghiri, M.: 'HDMAC-coding for MAC compatible broadcasting of HDTV signals, signal processing of HDTV', Proceedings of the 3rd International Workshop, Turin 30th Aug. – 1st Sept. 1989, pp. 187–194

151 HAGHIRI, M.: 'HDMAC: European proposal for MAC compatible broadcasting of HDTV signal', Philips Laboratoires, Limeil-Brévannes, France, IEEE 1990 CH2868-8/90/0000-1891

152 BERNARD, P., BLAIZE, C., and Colaitis, M.-J.: 'HDMAC coding scheme and compatibility', CCETT, Cesson-Sevingne Cedex, Signal Processing of HDTV. Proceedings of the 3rd International Workshop, Turin 30th Aug. – 1st Sept. 1989

153 MÄUSL, R.: 'HDMAC – a D2-MAC-comptabible HDTV transmission method', *Television technology, R&S refresher topics IX/12*, pp. 38–39

154 RICHTER, H.P.: 'Digitale Fernsehstudios für die Bildformate 4:3 und 16:9'. BTS Broadcast Television Systems GmbH, FKTG-Proceedings of 15th annual conference of FKTG, pp. 430–457

155 HOFMANN, H.: 'Verteilung digitaler Bildsignale im Fernsehstudio', *Fernseh- und Kinotechnik*, 1984, **38,** (11), pp. 477–482

156 HOFMANN, H., and KRÜGER B.: 'Programmaustausch im Bildformat 16:9 – Übertragung digitaler TV-Komponentensignale in Studioqualität über das vermittelnde Breitbandnetz (VBN) der DBP Telekom', *Rundfunktechnische Mitteilungen*, 1991, **35,** (6), pp. 272–277

157 NN: 'Endcoding parameters of digital television for studios', *CCIR: Recommendation 601-1*, ITU/CCIR: Recommendations and Reports of the CCIR, Geneva 1986, **XI,** Pt. 1, pp. 319–328

158 NN: 'Interfaces for digital component video signals in 525-line and 625-line television systems', *CCIR: Recommendation 656*, ITU/CCIR: Recommendations and Reports of the CCIR, Geneva 1986, **XI,** Pt. 1, pp. 346–358

159 NN: 'Transmission of component-coded digital television signals for contribution-quality applications at bit rates near 140 MBit/s', *CCIR: Recommendation 721*, ITU/CCIR: Recommendations and Reports of the CCIR, Geneva 1990, **XII,** pp. 68–79

160 HOFMANN, H.: 'Verteilung digitaler Bildsignale im Fernsehstudio', *Fernseh- und Kinotechnik*, 1984, **38,** (12) pp. 536– 546

161 BÜCKEN, R.: 'Digitalisierung der Studiotechnik macht Fortschritte – immer mehr Hersteller präsentieren digitale Kameras', *Fernseh- und Kinotechnik*, 1992, **46,** (10) pp. 704– 709

162 THIELE, M.: 'Im Studio beginnt die digitale Zukunft. Digitaler HDTV 1,2 GBit-Recorder', *Funkschau*, 1992, 23, pp. 48–51

163 DESOR, H.J., and Heberle, K.: 'Offene Systemarchitektur für digitale Signalverarbeitung', Reprint from *Elektronik*, 1991, **25,** No. 6200-224-1D, ITT Intermetall

164 FISCHER, T.: 'Schaltungstechnik eines Fernsehgerätes mit DIGIVISION', ITT Intermetall Freiburg, *Funk-Technik*, 1983, **38,** (5)

165 JORGENSEN, B.: 'Digital TV will pre-empt HDTV, spin off entirely new chip markets', *Electronic Business*, April 1993, p. 26

166 ERNST, H.: 'Digitale Bildverarbeitung – Grundlagen und technische Anwendungen', *Fachhochschule Rosenheim*, script for Rohde & Schwarz advanced training course DB 11, 9/88

167 SCHAMEL, G.: 'Codierung für die digitale terrestrische HDTV-übertragung', Heinrich-Hertz-Institut für Nachrichtentechnik Berlin GmbH, script of lecture held at 15th annual conference of FKTG in Berlin, June 1992

168 WOOD, D.: 'How much bit-rate reduction is possible? European perspectives', *Montreux International Television Symposium*, 1991, pp. 601– 611

169 HARTWIG, S., and ENDEMANN, W.: 'Datenkompression als Schlüsseltechnik der digitalen Bildverarbeitung', Tutorial Digitale Bildcodierung, *Fernseh- und Kinotechnik*, 1992, **46,** Part 1, (1), pp. 23–30

170 KORN, A.: 'Bildverarbeitung durch das visuelle System' (Springer, Berlin Heidelberg New York; 1982)

171 HARTWIG, S., and ENDEMANN, W.: 'Das menschliche visuelle System als Nachrichtensenke bei der datenkomprimierten Bildübertragung', Tutorial Digitale Bildcodierung, *Fernseh- und Kinotechnik*, 1992, **46,** Part 2, (2), pp. 112 –118

172 HARTWIG, S., and ENDEMANN, W.: 'Das Bild als Nachricht Tutorial Digitale Bildcodierung', Fernseh- und Kinotechnik, 1992, **46,** Part 3, 3, pp. 187–192

173 HARTWIG, S., and ENDEMANN, W.: 'Dekorrelation und Quantisierung', Tutorial Digitale Bildcodierung, Fernseh- und Kinotechnik, 1992, **46,** Part 4, (4), pp. 249–254

174 HARTWIG, S., and ENDEMANN, W.: 'Codierung von Fernsehbildern mit der DPCM', Tutorial Digitale Bildcodierung, *Fernseh- und Kinotechnik,* 1992, **46,** Part 5, (5), pp. 334— 342

175 HARTWIG, S., and ENDEMANN, W.: 'Bewegungkompensierte Interframe-DPCM', Tutorial Digitale Bildcodierung, *Fernseh- und Kinotechnik,* 1992, **46,** Part 6, (6), pp. 416 –424

176 MUSMANN, H.G., PIRSCH, P., and Grallert, H.-J.: 'Advances in picture coding', *Proceedings of IEEE,* April 1985

177 HARTWIG, S., and ENDEMANN, W.: 'Transformationscodierung (1)', Tutorial Digitale Bildcodierung, *Fernseh- und Kinotechnik,* 1992, **46,** Part 7, (7–8), pp. 505– 511

178 GILGE, M.: 'Regionenorientierte Transformationscodierung in der Bildkommunikation', VDI-Verlag, Fortschrittsberichte, Reihe 10, 1990

179 HARTWIG, S., and ENDEMANN, W.: 'Transformationscodierung (2)', Tutorial Digitale Bildcodierung, *Fernseh- und Kinotechnik,* 1992, **46,** Part 8, 9, pp. 597– 605

180 HARTWIG, S.; and ENDEMANN, W.: 'Digitales Fernsehen: Vom Labor zum Standard', Tutorial Digitale Bildcodierung, *Fernseh- und Kinotechnik,* 1992, **46,** Part 9, (10), pp. 687–696

181 HARTWIG, S., and ENDEMANN, W.: 'Vektorquantisierung', Tutorial Digitale Bildcodierung, *Ferseh- und Kinotechnik,*1992, **46,** Part 10, (11), pp. 763 –775

182 PARKE/MORRIS: 'International standards for digital TV coding', *IEE Colloquium on Prospects for Digital Television Broadcasting No. 137,* London BD 137 1991, pp. 3/1–5

183 NN: 'Bild- und Video-Kompression nach JPEG- und MPEG-Verfahren – als Single-Chip-Lösung', M*etronik/C-cube* M*icrosystems,* data sheet May/June 1992

184 HEPPER, D.: 'ISO-MPEG – Der Beginn der Standardisierung digitaler Bildcodierung im Konsumentenbereich', *Fernseh- und Kinotechnik,* 1992, **46,** (4) pp. 238ff

185 HARTWIG, S., and ENDEMANN, W.: 'Teilbandcodierung', Tutorial Digitale Bildcodierung, *Fernseh- und Kinotechnik,* 1992, **46,** Part 11,(12), pp. 846–856

186 HARTWIG, S., and ENDEMANN, W.: 'Von Signalen zu Objekten - Neue Wege in der Bildcodierung', Tutorial Digitale Bildcodierung, *Fernseh- und Kinotechnik,* **47,** 1993, Part 12, (1), pp. 33–42

187 PLANSKY, H., SCHIEFER, P., and RUGE, J.: 'Bilddatenreduktion und ihre Anwendung bei HDTV', *Frequenz,* 1992, **46,** (3–4), pp. 102–109, Lehrstuhl für Integrierte Schaltungen der Technischen Universität München

188 LIEBSCH, W.; TALMI, M., and WOLF, S.: 'Prozessoren für die Echtzeit-Bildtransformation von HDTV-Signalen', Heinrich-Hertz-Institut für Nachrichtentechnik Berlin GmbH, pp. 175–180

189 HEPPER, D.: 'Digitale terrestrische HDTV-Übertragung – Probleme und Lösungen', Script of lecture held at 15th annual conference of FKTG in Berlin, June 1992

190 OSTROM, C.: 'Digital video compression – the basic concepts', NAB 1992, Broadcast Engineering Conference Proceedings, pp. 345–353

191 MUSMANN, H.G.: 'Entwicklung der digitalen Ton- und Bildcodierung', *Forum 'Innovatives Europa',* Bonn – Bad Godesberg, March 1990

192 WEISS, P., CHRISTENSSON, B., ARVIDSSON, J., and ANDERSSON, H.: 'Design of an HDTV codec for terrestrial transmission at 25 MBit/s', *HD-*

DIVINE Project, pp. 14–17, Vistek Electronics, UK
193 NN: 'Digital television broadcasting', EBU/ETSI JTC 6(92)30 6th Meeting, Sophia Antipolis, October 1992
194 HELLER, J., and PAIK, W.H.: 'The digiCipher HDTV broadcast system', *Montreux Record*, 1991, pp. 595–600, VideoCipher Division, General Instrument Corporation, USA, California
195 LOOKABAUGH, T.: 'Compressed digital video: a technology overview', Broadcast Engineering Conference Proceedings Compression Labs, Inc., San Jose, California 408-NAB 1992,
196 NN: '1125/60 high definition origination in the era of conversion to advanced television', pp. 3–14, HDTV 1125/60 Group 1615 L St. NW Washington DC,
197 MILLER, M.: 'Digital HDTV on cable', Vice President, Technology and New Business Development, Jerrold Communications, 2200 Byberry Road, Hatboro, PA, USA 19140, *Montreux Record*, 1991, pp. 427–432
198 KELKAR, K.: 'DigiCipher: an update on progress made in the United States,' pp. 375–379,General Instrument Corporation, USA
199 LUPLOW, W., and FOCKENS, P.: 'The all-digital spectrum-compatible HDTV system', Zenith Electronics Corporation, Glenview, IL. 60025 USA, *Montreux Record*, 1991, pp. 169–184
200 STARE, E.: 'Development of a prototype system for digital terrestrial HDTV', Reprint from *Tele, English Editions No. 1/92, Tele 2/92*, pp. 1–6, Telia Research AB
201 BERNARD, P.: 'Sterne: the CCETT proposal for digital television broadcasting', CCETT, France, pp. 372–374
202 MONNIER, R., RAULT, J.B., and DE COUASNON, T.: 'Digital television broadcasting with high spectral efficiency', Thomson-CSF/LER, France, pp. 380–384
203 MASON, A.G., and LODGE, N.K.: 'Digital terrestrial television', development in the spectre project', National Transcommunications Ltd. & Independent Television Commission, UK, Proceedings IBC 92, July 1992 Amsterdam, IEE Conf. 358, pp. 86–91
204 LONG,T.: 'Digital television broadcasting developments in Europe', Independent Television Commission, Winchester, Hampshire, United Kingdom, NAB 1992 Broadcast Engineering Conference Proceedings, pp. 126–135
205 NN: 'HDTV-T digitale terrestrische HDTV-übertragung', Description of joint project
206 MÜLLER-RÖMER, F.: 'HDTV: der Sprung ins nächste Jahrhundert', *Funkschau*, 1993, **8**, pp. 49–53
207 DAMBACHER, P.: 'The way to DAB via digital satellite radio DSR', Montreux 1991, Conference Proceedings
208 DAMBACHER, P.: '1. Der Weg zu DAB über Digitalen Satelliten-Rundfunk, 2. Ein DSR/MUSICAM-verfahren mit z.B. vierfacher programmkapazität', 2ARDA, August 1991, special print in German
209 ASSMUS, U.: 'Datenübertragung im DSR', *Rundfunktechnische Mitteilung* 1989, Issue 1
210 MAIER, J.: 'Digitalfunk/Im Brennpunkt', *Stereoplay*, 11, 1989
211 NN: SAA 7500, preliminary data sheet, Valvo 4/88
212 NN: 'Digitaler Hörfunk am Scheideweg', Funkschau, 1990, **2**
213 DAMBACHER, P.: 'Verfahren und Einrichtung zum Übertragen der digitalen Tonsignale von Tonstudios zu den einzelnen Sendestationen eines Rundfunk-Sendernetzes', Rohde & Schwarz Patent Specification 1251-P

214 NN: 'Access control system for the MAC/packet-family', *Eurocrypt, Final Draft prEN 50094*, August 1991, CENELEC
215 DAMBACHER, P.: 'Finding the path to DAB', *Radio World International*, 1991, **15**, (24), p. 27 (and following issues)
216 SCHNEEBERGER, G.: 'Verteilung von DSR-Signalen in 12-MHz-Kanälen', Institut für Rundfunktechnik, March 1987, MVSB/3SCH
217 NN: 'Einspeisung eines weiteren DSR-Paketes in das BK-Netz München', *Rohde & Schwarz Technical Information*, April 1993
218 NN:'Dem DSR gehört die Zukunft', *Infosat*, 1992, **Issue 11,** (56), pp. 32–34 px11/32sc
219 NN: 'Die digitale Radiovielfalt – DMX in den USA erfolgreich – DSR-Plus startet für Europa', *Infosat*, 1992, **Issue 12,** (57), pp. 118–119 (mwo)
220 MÄUSL, R.: 'Digitale Modulationsverfahren', Hüthig Verlag, pp. 186 and 244
221 NN: 'Satellite tuner FT990', Preliminary data sheet from Philips
222 NN: 'VADIS', Report of the third EBU/ETSI Joint Technical Committee (JTC) Meeting in Geneva on 11th–12th February 1991, Item 5.7
223 DAMBACHER, P.: 'Verfahren zum Übertragen von digitalen HDTV-Signalen, 2AR-Da', Rohde & Schwarz special print, August 1991
224 DAMBACHER, P.: 'Verfahren zum Übertragen von digitalen HDTV-Signalen'. Rohde & Schwarz employee invention 1270-P, P4125 606.0
225 MÄUSL, R.: 'Untersuchung zur Bitfehlerwahrscheinlichkeit pe bei einer 8PSK/2AM im Vergleich zu 16QAM', Study of 6.1.1993 in co-operation with P. Dambacher
226 DAMBACHER, P.: 'Digitales Rundfunksendernetzsystem'. Rohde & Schwarz patent specification with supplement and enhancement of employee invention P4231536.0 (1297-P)
227 BAUER, H.: 'Erzeugung von 4PSK- und 8PSK/2AM-Signalen'. Rohde & Schwarz study, January 1993, in co-operation with P. Dambacher
228 BAUER, H.: 'Die Modulation 16QAM und 8PSK/2AM bei nichtlinearer Übertragungskennlinie der Wanderfeld-Röhre'. Rohde & Schwarz study, February 1993, in co-operation with P. Dambacher
229 STARK, A.: 'Untersuchung der unterschiedlichen Laufzeitverzögerung bei Programmzuführung zu DAB-Sendern via Satellit'. Rohde & Schwarz study of 6.8.1992, in co-operation with P. Dambacher
230 DIETL, A., and KLEINE, G.: 'SFP, DSRU and DSRE – the pleasure of digital sound with CD quality', *News from Rohde & Schwarz*, 1991, **135**
231 DAMBACHER, P.:'Programm and data lines to VHF-FM-transmitters', Rohde & Schwarz study 2AR-Da 10.90
232 BOCHMANN, H.: 'Vier Antennen an einem Empfänger', *Funkschau*, 1992, 1, pp. 66–70
233 LINDENMEIER, H., and REITER, L.: 'Möglichkeiten und Grenzen des UKW-Empfanges mit Antennen-Diversity im Kraftfahrzeug', *Rundfunktechnische Mitteilung*, 1984, **II**, 28
234 BAIER, P.: 'Raum-Diversity-Betrieb für VHF/UHF-Relais', *Funkschau*, 1992, **12,** pp. 74–77
235 LOHSE, K.: 'Antennen-Diversity, Besserer Radio-Empfang im fahrenden Auto', *Funkschau*, 1993, 4, pp. 56– 59
236 KUHL, K.H., PETTKE, G., SCHNEEBERGER, G.: 'Zur Gestaltung von Sendernetzen für einen digitalen terrestrischen Hörrundfunk', IRT, 8.10.90
237 UNGERBÖCK.: 'Trellis-coded modulation with redundant signal set', *IEEE Communications Magazine*, 1987, **25,** (2)

238 SCHNEEBERGER, G.: 'Digitales Fernsehen – Warum und Wie'. IRT study, January 1990, Digvid revl

239 NN: 'LCD – Farbfernsehgeräte', *Funkschau*, 1992, 18, pp. 13–22

240 NN: 'Plasma-Display im HDTV-Breitbild-Format'. Genschow Technischer Informationsdienst, Auslandschnellbericht 33 – 1992, Ausgabe B, pp. 2–3

241 FLEMING, M.E.: 'An ultra-high speed DSP chip set for real-time applications'. Sharp Electronics, Hamburg, data sheet issued April 1993

242 NN: 'Optische Speichersysteme, hohe Speicherkapazitäten und Datensicherheit', *NET 41*, 1987, Issue 6, pp. 233–234

243 DAMBACHER, P.: 'Verfahren zur digitalen Modulation mit Sinusträger'. Rohde & Schwarz Patent Specification P43 18 547.9 (1362-P)

244 FISCHER, W.: 'Die Fast-Fourier-Transformation – für die Videomeßtechnik wiederentdeckt', *Fernseh- und Kinotechnik*, 1988, **42**, (5) pp. 201–208

245 HEINEMANN, C.: 'COFDM mit digitaler I-Q-modulation'. Rohde & Schwarz study of 18th October 1993, in co-operation with P. Dambacher

246 ZSCHUNKE, W.: OFDM-Übertragung mit konstanter Hüllkurve'. Technical University Darmstadt, Brief information dated 25th October 1993

247 NN: 'SES-Digitales Fernsehen auf Astra'. MPM/93-F142 E.2 to E.6

248 NN: 'The European digital video broadcasting DVB', Baseline System, Baseline Specification. The MULTIVISION System EBU 6.1.1994 DT/8038/III-B ld

249 NN: 'Amerika bekommt jetzt digitales Fernsehen'. Blick durch die Wirtschaft, 29th December 1993

Index